Petroleum Lands and Leasing

PETROLEUM LANDS and LEASING

Joan Burk

PennWell Books
PennWell Publishing Company
Tulsa, Oklahoma

PennWell Publishing Company
1421 South Sheridan Road/P. O. Box 1260
Tulsa, Oklahoma 74101

Library of Congress cataloging in publication data

Burk, Joan.
Petroleum lands and leasing.

Bibliography: p. 163
Includes index.
1. Petroleum law and legislation—United States. 2. Oil and gas leases—United States. 3. Real property—United States. I. Title
KF1849.B77 1983 346.7304'68232 83-7283
ISBN 0-87814-239-8 347.3064 68232

Printed in the United States of America

2 3 4 5 87 86 85 84

To **Bryan Burk,** CPL
Oil and Gas Consultant
Friend and Lover

Contents

Preface

This book is intended as a helpful reference to all who are interested in petroleum lands and leasing, as lessor, lessee, or landman. It is a compilation of information obtained from research and experience.

Although legal citations are referenced for those who may wish to pursue a particular subject in detail, the text does not purport to give legal advice or legal interpretations. Federal regulations and state laws are cited to illustrate general principles.

The data included in this book will assist the reader to understand and to negotiate oil and gas properties situated in the domestic United States and Alaska. The interests covered are inland, onshore lands of private, federal, state, and Indian ownership.

Petroleum lands are comprehensive. Therefore, this book must be restricted in some manner. It does not deal with gas purchase contracts nor the producing aspects of petroleum properties.

Gratitude and sincere appreciation are extended to many who have assisted me through the years and with the compilation of information for this book. Those who gave me early training and experience in Dallas, Texas, include Vernon C. Howell, independent oil operator; and E.B. Germany and Sons, without whose helpful guidance and offers of opportunity I would not be a landman. The American Association of Petroleum Landmen is a superb organization that has given me the opportunity to serve in capacities that have allowed me to meet the leaders of our industry, leaders from whom I have learned in each encounter. Friends in the land and legal departments of Exxon Co., U.S.A., offered me learning experiences of lasting value.

In particular, attorney at law, Candace Weiss, contributed to this book by teaching me methods of legal research and by looking over my shoulder to prevent errors common to the nonattorney. Gratitude is also expressed to my editor, Kathryne E. Pile, for her faith in me.

For any accomplishment I may attain, I owe a debt of appreciation to my mother, Sylvia Ricketts, who has been the encouragement and inspiration of my life; to my kind and supportive husband, Bryan Burk; and to the good Lord, who makes all things possible.

Joan Burk, CPL

CHAPTER ONE

A Historical Background

Title to land by an individual extends back to Roman tradition under the Justinian Code of the sixth century. This code, known as *Corpus Juris,* became the basis of the Spanish code of law that later governed royal land grants in America.

New England's history is familiar. British settlers brought with them the English common law that was adopted by the framers of the Constitution of the United States. The English common law is based on decisions made by judges. It originated in the twelfth century and continues to develop according to the rule of precedent, or *stare decisis,* which means "let the decision stand." If a rule is established by a court, the common law requires that the same rule shall be followed in subsequent similar cases. Each state has its own constitution based on the English common law with the exception of Louisiana, where the law is based on the Napoleonic Code (or civil law).

Rulings may also be based on equity. A person is entitled to show why the common law remedy might be inadequate and ask for equitable relief from the court.[2]

Sovereignty of the Soil

In a counterbalance of authority and diplomacy, Pope Alexander VI drew an imaginary line to settle land rights between two Catholic powers, Spain and Portugal. This action followed the return of Columbus from his first voyage in 1493. Apportionment of the primitive worlds by papal bull created the first claims of sovereign title in the Americas. Pope Alexander's line ran north-south 100 leagues west of the Azores and Cape Verde Islands, skimming the mainland coast of the South American continent. Portugal gained newly discovered lands east of the line; Spain could claim lands lying west of the line.

Neither nation was content with Rome's settlement of its territorial disputes. The line of demarcation was reset on June 7, 1494, under the Treaty of Tordesillas between Castile and Portugal. This line was never surveyed, but authorities place it between 48° and 49° W. Long. The intent was to extend the line of demarcation all the way around the globe. By that agreement, Spain acquired title to "New Spain" (in the Americas and the Indies). However, England, France, and The Netherlands never recognized the jurisdiction of the pope nor the Treaty of Tordesillas with regard to their own exploration of the areas.

Spain's newly discovered lands were not considered Spanish colonies per se. The lands constituted a new group of kingdoms under the Royal Council of Castile. Columbus made his fourth and last voyage during the years 1502–1504. His career ended with no grants of land. None of the lands he discovered even perpetuated his name. Charles Gibson, author of *Spain in America,* describes the "Vespucci question" aptly. Quoting Clements Markham, the editor and translator of Amerigo Vespucci's letters, Gibson tells us that

> The evidence against Vespucci is cumulative and quite conclusive. His first voyage is a fabrication. He cannot be acquitted of the intention of appropriating for himself the glory of having first discovered the mainland.[1]

This is not to debate who first touched the North American mainland. Amerigo prevailed, for it is his name that graces the Americas.

Finding the lands already populated, the Spanish conquerors faced problems of conscience. They eventually prepared a lengthy proclamation called the *Requermiento* (Requirement). This proclamation was to be read aloud to tribes of Indians encountered by each conquistador, in the presence of an interpreter, a notary, and proper witnesses. Thus the conscience of the Spaniard was absolved as he read the Indian heathen a detailed account of the authority of the Catholic Church, the pope, and the monarch. All who failed to swear allegiance fell under due consequences and punishment.

Hernándo Cortés was the largest and richest landowner under the encomendero system. An encomendero was a patron to whom was granted the use of lands and the right to use the Indian natives upon it for labor. In turn, he agreed to be responsible for the welfare of his charges. Demands for rights of inheritance began with Cortés.[3] However, no land titles were actually granted in the name of any of the conquistadores. Perpetuity of ownership in Spanish land grants was finally established under the Ordinances of Segovia of 1573.

New Mexico was colonized and in 1573 received its official name from Juan de Oñate. Oñate also claimed for Spain the territories now called Arizona, Colorado, Nevada, Texas, Oklahoma, and Kansas. He traveled across the Arizona territory during 1605 and 1606. When he found California, Oñate proclaimed it to be an island. That belief persisted for a hundred years or more. Junipero Serra founded the first California mission in 1769, in San Diego. The Jesuits moved northward along the coast.

The French and Spanish were bitter rivals for the southeastern Gulf Coast lands. French explorers descended the Mississippi River in 1673 to establish a fort at Biloxi, Mississippi. They were the first large group of white men to move into the Alabama area in 1702. Meanwhile, British frontiersmen moved from the Tennessee Valley southward into northern Alabama. They called it "Illinois Country" and set the area aside for Indian hunting grounds. France was forced to cede its North American colonies to Britian after the French and Indian War according to the Treaty of Paris of 1783.

Spain joined the American Revolutionary War against Britain. Britain relinquished the Mobile area to Spain in a second Treaty of Paris, 1783, and Spain emerged holding the entire southern half of the present United States. She encountered a scarcity of civilized population but allowed no foreigners in to settle it. Unable to govern such vast lands, Spain ceded the Louisiana territory to France in 1802. Much to Spain's chagrin, France made the Louisiana Purchase available to the United States in 1803.

The United States moved southward and seized Mobile during the War of 1812. Seven years later, Spain ceded Florida to the United States. The Treaty of San Lorenzo fixed the southern boundary of the United States on the 31st parallel.

Spain was gripped by paranoia that French or English settlers would wrest its territories away from it. For this reason Spain could never bring itself to allow outsiders permission to settle the expanse of its southwest stronghold. Mexico's declaration of independence from Spain in 1821 was the beginning of the end of Spanish sovereignty in North America. Mexico encouraged American settlers to move into Texas with promises of "a league and a labor" for 12½¢ an acre, under the empressario system begun by Moses and Stephen F. Austin.

Texas declared itself a republic in 1836 and entered the Union in 1845. Texas brought with it unique circumstances in land history and law from having been the only sovereign country within the continental boundaries of the United States. Texas adopted the English common law but retained vestiges of Spanish legal traditions that survive today.

The Treaty of Guadalupe Hidalgo in 1848 gave the United States all of the lands claimed by Mexico from the Texas border westward to California. "Every deed to property along the Rio Grande in Southwest Texas had its origin in a Spanish porción, adjudicated when the Rio Grande Valley was part of Spain."[4] Miller v. Letzerich, 121 Tex. 248, established the validity and legal effect of contracts and grants in effect before Texas' sovereignty.

Virginia H. Taylor relates a true event in the preface to her *Index to Spanish and Mexican Land Grants.* The first step taken by the new Republic of Texas was to accept as valid and approve previous Spanish and Mexican grants to lands in the Texas territory. (This policy carried into statehood after 1845.) The Bourland and Miller Commission was established to investigate title claims to grants in the area situated between the Nueces and Rio Grande rivers. James B. Miller made a trip to the Rio Grande Valley and Mexico to collect original deeds that he could take to Austin for validation and the issuing of patents. He left Brownsville in November 1850 on the steamer *Anson* with dozens of original grants packed in a trunk. Unfortunately, the ship encountered a storm and began filling with water. It broke apart and sank only 200 yards from shore, near Matagorda Island. Although Miller was saved, the original grants sank with the trunk into the turbulent waters of the Gulf.

The Spanish were excellent and meticulous record keepers, but for some reason they never adopted an index system to catalog their grants. Claims to Spanish and Mexican land grants continue even today in the courts of Texas and New Mexico.

Under the expansion of the United States, the Homestead Act of 1862 was adopted. Individuals could acquire 160 acres for a residence by occupying the land for five years and paying small fees. These patents were not as generous as the Spanish league (4,428 acres) but were more in accord with the Spanish labor (177 acres) that provided land for homes and crops.

A treaty line established in 1818 set the U.S.-Canadian border. It gave the United States a large territory north of Louisiana that the British had called the Red River Basin. U.S. Secretary of State William Seward purchased Alaska from the Russians on March 30,

1867. Amazing as it may seem, Congress did not provide for an Alaskan government for 17 years. A civil code was finally adopted for Alaska in 1900. A border dispute between Canada and Alaska was settled by an international commission in favor of the United States. Alaska adopted its present state constitution in 1956, three years prior to statehood. Ninety percent of Alaska's land continues to be state and federally owned. Only 10% is privately owned or administered by local governments in organized boroughs.

Thus the lands of the United States came together under sovereignty of the federal government and the states. The last fragment of the continental United States was acquired in 1853 with the Gadsden Purchase. This was a strip of land that the United States bought from Mexico to settle a dispute of ownership along the southern edge of Arizona and New Mexico. In 1979, Governor William Clements of Texas deeded a small piece of land (called the Chamizal) along the Rio Grande River to Mexico in order to clear a border dispute.

The granting of land patents by state and federal authorities for private ownership of the surface of land and the mineral wealth below it is a special privilege enjoyed by citizens of few countries. Recognizing the divergent history affecting our state laws and land titles brings a better understanding of the legal problems and decisions that affect our development of petroleum resources.

Chronology of the Oil Industry in the United States

In July 1543, remnants of Hernándo de Soto's expedition straggled across Texas toward the Gulf of Mexico. The men arrived at an area generally thought to be about three miles west of Sabine Pass. They stopped in the swampy waters to repair their portage boats with a pitchlike substance found on the surface of the inland rivulets.[5]

Earliest stories of the history of petroleum products in the United States come from the journals of French missionaries. These men used a petroleum liquid skimmed off the surface of oily springs near a town now known as Cuba, New York. Jesuit missionaries found "a thick, oily stagnant water that would burn like brandy" near Lake Erie about 1670.[6] Oily liquids from springs in the Nacogdoches area of Texas were used by the Indians and white inhabitants for medicinal and lubricating purposes.[7]

Mass killings of whales between 1830 and 1850 created a scarcity of whale oil, which had been the common oil used for illumination, and the United States soon outgrew the available supply. Whale oil sold for $2.00 to $2.50 per gallon by 1850. Oils from other animals proved unsuccessful substitutes. Alcohol, turpentine, paraffin oil, and other mixtures were tried in variously designed lamps. Obviously, a new source of illuminant had to be discovered.

Drilling for petroleum evolved from drilling for salt, a common practice in West Virginia before the end of the eighteenth century. Petroleum seepage and natural gas deposits interfered with the mining of salt.

An 1826 magazine article mentioned burning gas springs in the Kanawha River Valley, West Virginia, in connection with a salt well on the Little Muskingum River. "It discharges such vast quantities of

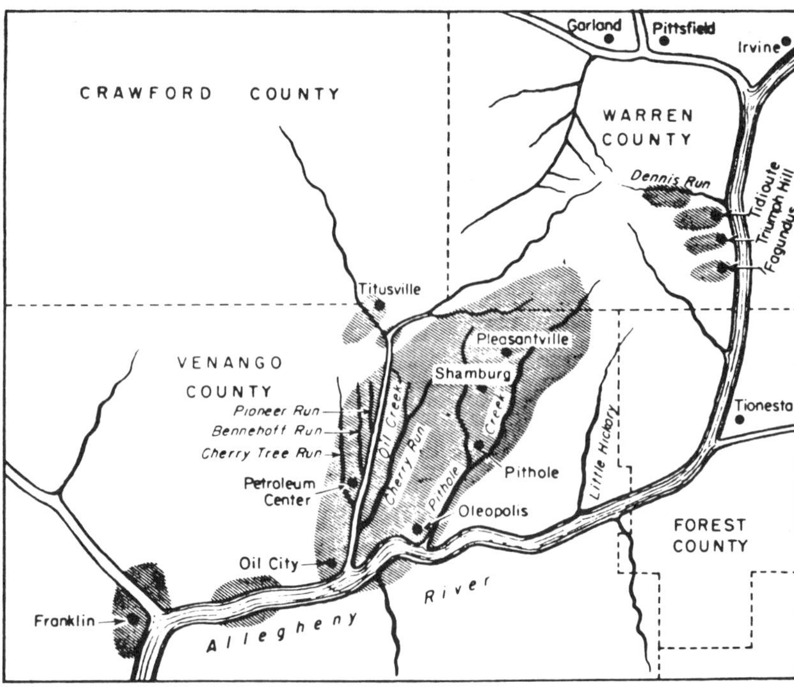

Fig. 1–1 Upper Region oil-producing territories, 1860s (reprinted from Williamson and Daum, *The American Petroleum Industry* (Evanston, Ill.: Northwestern University Press), p. 129, with permission of Northwestern University Press.)

petroleum and is subject to explosions of gas as to force out all the water, that it makes little or no salt. It makes a bright light that might be suitable for lighting the street lamps of the cities of Ohio."[8]

The greatest visible source of petroleum in the United States prior to 1846 was along Oil Creek in northwestern Pennsylvania. Settlers followed a custom that the Indians had used for years. They skimmed off the oil and used it for medicines. An enterprising man from Pittsburgh named Samuel Kier decided to bottle and sell the petroleum recovered near his father's salt wells near Tarentum, Pennsylvania. Kier distilled the oil to help eliminate unpleasant odors and smoke. He called his product "carbon oil." It soon provided a decent source for illumination purposes, and Kier's price rose from 75¢ to $2.00 a gallon.

Abraham Gesner, a geologist from Nova Scotia, made oil from Albert coal in New Brunswick. Subsequently, he found he could make oil from asphalt. He called his new product "kerosene," a contraction of the Greek words for "coal" and "oil."[9] J. H. Williams in Ontario drilled a shaft to 60 feet in gravel and brought a heavy oil to the surface by pump. These inventive measures preceded the first well of any note in the United States.

News spread rapidly when E. L. Drake brought in his well at Titusville, Pennsylvania, on August 29, 1859, yielding 20 barrels of oil a day. The hunt for oil had begun. Cable tool drilling, a method developed for water and salt drilling by the Chinese prior to 1700, was the common practice.[10] By 1869, 5,000 wells had been drilled in the United States. About one-fourth of these were productive, yielding 25 million barrels of oil by 1870.

Kansas saw its first good well drilled in 1860 in Miami County, approximately 40 miles south of Kansas City. The Civil War lured the drillers to the fight, and the search for oil ceased temporarily. Also in 1860, A.M. Cassidy drilled a test well in an area near what is known today as Cañon City, Colorado. His first well was only 24 feet deep and produced one barrel of oil per day. Cassidy drilled other wells and produced oil from shale formations, a forerunner of the present Florence field.

In Texas, Lynis T. Barrett emigrated toward Nacogdoches County to find the oil seepages he had heard about back in his home in Virginia. When the Civil War ended, Barrett drilled the first successful well in Texas, using the rotary drilling method.[11] It produced about 10 barrels of oil per day, compared to Drake's 20 barrels. Another 20 years would pass before Texas would have a significant field.

Leasing procedures varied in early 1860s. Rapid development created close spacing with little or no regard for drainage or conservation. Wasteful practices brought shortlived wells. Landowners leased in smaller and smaller tracts to obtain more wells. Prudent operators saw the problems increasing. Lessees sought to remedy matters in new leases by specifying how many wells could be drilled on the acreage. They inserted restrictions on the withdrawal of tubing and required abandoned wells to be plugged.[12]

Leases tended to protect the landowner. They frequently stipulated that the lessee had to drill a well by a certain time and had to pursue his operations "with due diligence." The standard one-eighth royalty did not come into effect until about 1870. Some early royalties were as low as one-tenth or one-twelfth, but those were in the minority. Royalties during the Oil City, Pennsylvania, boom ran from one-fourth to three-fourths. Fractional working interests in leases were as small as 1/128. The selling of fractional shares of oil and gas leases provided capital for the promoter and lessened his risk.

Larger blocks began to be leased, and the lessee began to pay annual delay rentals. Primary terms varied from 20 years to 99 years, even to perpetuity. Usually a specified time—six months to a year—was set for the commencement of drilling, and failure to perform required cash payments. The use of the cash payment as an up-front bonus began about 1861. Sometimes it was paid when the lease was executed, but frequently the sum would be payable on the contingency of production in certain quantities.[13]

The first West Coast production took place in Humboldt County, California. Drilling moved southward toward Los Angeles. The Summerland pool, which was situated offshore, comprised California's first sizable field. In 1895, Summerland had 28 producing wells; by 1900, it had 305.[14] Production began in Santa Barbara County, then in Los Angeles County, near the La Brea Tar Pits. City sites were eventually abandoned, however. Real estate in Hollywood boomed, and land was worth more for its surface use than for its oil potential.

Few people who became active in the oil business had any training in geology or minerals. Oilmen gradually realized the significance of knowing about the rocks that lay below the surface. The predominant belief was that if oil was found on the surface of water in a creek, that was the place to drill. Finally, the realization evolved that the difference between drilling in valleys and on hills was that deeper drilling would be required from higher elevations.

Years before the geology of the area was understood, the Cincinnati Arch (extending from Indiana into Ohio) produced prolific gas from the Lima-Indiana field.

Gasoline was refined as early as 1865, before its principal use was invented. New means of using petroleum products were constantly sought. Charles Duryea built a gasoline horseless carriage in 1892, and Henry Ford developed his four-wheel automotive vehicle.

Production came to Oklahoma with small amounts of oil from a 120-foot well that was drilled in what is now northeast Rogers County, on Cherokee Nation lands. Even then, "historians in Oklahoma . . . stressed the difficulties of securing valid leases on Indian lands . . . due to the uncertainty of the rights of tribal councils or officials to make the leases."[15]

The drilling of a water well brought in oil at Corsicana, Navarro County, Texas, at 1,027 feet. This lucrative field has continued to produce since 1894. Spindletop's gusher of 1901 in Jefferson County, Texas is considered the most important well in early U.S. history. Spindletop proved that petroleum could be found on or around salt domes, and it opened the Gulf Coast as a major source of petroleum production.[16]

Legal entanglements increased with the leasing and use of land for petroleum development. Courts based decisions on theories of property law from which the rule of capture evolved, citing an English doctrine that dealt with percolating waters, Acton v. Blundell, 12 Meeson & Welsby 324, 1843. The rule of capture recognized oil and gas as migrant and declared that the surface owner had the right to take all oil and gas that could be obtained from under his land.

As noted earlier, even in the 1800s, oil and gas leases protected the lessor with clauses that required the pursuit of operations "with due diligence." The same diligence and prudent operations formed a basis for the standards attained by the oil industry in the United States during its 125 years of phenomenal growth.

References

[1]Charles Gibson, *Spain in America* (New York: Harper & Row, 1966), footnotes 17 and 18, pp. 13, 14.

[2]James MacGregor Burns and Jack Walter Peltason, *Government by the People* (Englewood Cliffs, N. J.: Prentice-Hall), pp. 434 ff.

[3]Gibson, p. 55.

[4]Betty Eakle Dobkins, *The Spanish Element in Texas Water Law* (Austin: University of Texas Press, 1959), pp. 28–125.

[5]Paul H. Giddens, *Early Days of Oil* (Princeton, N.J.: Princeton University Press, 1948), p. 10.

[6]Harold F. Williamson and Arnold R. Daum, *The American Petroleum Industry* (Evanston, Ill.: Northwestern University Press, 1959), p. 10.

[7]Giddens, p. 1.

[8]Samuel W. Tait, Jr., *The Wildcatters* (Princeton, N.J.: Princeton University Press, 1946), p. 46.

[9]Williamson and Daum, p. 3.

[10]Tait, p. 46.

[11]Ibid., p. 43.

[12]Williamson and Daum, p. 161.

[13]Ibid., p. 92.

[14]Tait, p. 112.

[15]Ibid., p. 106.

[16]Ibid., p. 113.

CHAPTER TWO

Rights of Ownership

Ownership of real property in America has been indicated by the public recording of instruments of conveyance since the late eighteenth century. Prior to that time, in the territories of the South and Southwest owned by Spain, ceremonies of possession recognized the legal transfer of land to individuals. A settler explored an area, selected his tract, and applied to the land commissioner for title. When title was approved and the land was determined to be unclaimed, the new owner—in the presence of the land commissioner and neighbors—would walk upon the grounds, pick up stones and throw them, pull weeds, break sticks, and yell in a loud voice that the land was his. The presence of his neighbors at the ceremony assured that boundaries were agreed upon and did not overlap. The Spanish were meticulous record keepers. Deeds and certified copies were in common use prior to the occupation of North America. An official seal was placed on an instrument, and the signature of the seller or transferor included a flourish called a "rubric." The rubric was an identifying mark of the signature party, supposedly impossible to forge. Apparently, however, no indices of the many land transactions were compiled. Actual deeds were frequently lost, leaving the owner with no proof of his title. The old Western melodramas of the "stolen" deed are more fact than fiction.

Since the days of the Norman Conquest in 1066, the English common law has allowed grants of land to individuals. The earliest grants were given for services to the monarchy, usually in battle. A ceremony to transfer ownership was called a "feoffment." It differed only slightly from the ceremony observed by the Spanish by being more sophisticated. The seller of the land (or official of the sovereign) would take the grantee to the site and walk upon the land with him. A handful of dirt would be swept up and poured into the hand of the new owner, indicating the transfer of title. No written instrument was necessary.

Joint tenancies, remaindermen, undivided interests, and rights of inheritance became legal. Gradually, in order to effect a conveyance, the deed under seal became a requirement of the common law.

Theories of Ownership

The migrant nature of oil and gas has elicited two basic theories of ownership of these hydrocarbon products. Various states, including Alabama, California, Illinois, Indiana, Kentucky, Louisiana, New York, Ohio, and Wyoming, have adopted the nonownership or qualified-ownership theory. The theory assumes that oil and gas are fugitive and migratory, not owned until "captured" or produced. This is also referred to as an "incorporeal" status.

The "corporeal" or ownership-in-place theory grants the same nature to oil and gas as it does to solid substances. This theory is generally accepted by Arkansas, Colorado, Kansas, Maryland, Michigan, Mississippi, Montana, New Mexico, North Dakota, Pennsylvania, Tennessee, Texas, Washington, and West Virginia.[1]

A factor distinguishing the two theories is whether the landowner acquires title to oil and gas produced from his lands, even if the product has "migrated" from an adjoining tract. The migratory nature is the basic premise of the rule of capture, which allows an owner all of the oil and gas he may produce from his lands.

Following the explorations of Columbus and years of worldwide expansion and colonialism by the European nations, entire ocean areas were claimed by sovereigns. Today the United States government claims lands bordering the oceans and extending out to the edge of the continental shelf, a distance that varies from 100 miles into the Atlantic to 140 miles into the Gulf of Mexico and only 25 miles into the Pacific.[2] Actual ownership of the ocean areas has not been established under international law.

Fee Simple Estate

Fee simple ownership (also called fee or fee simple absolute) holds full right, title, and interest to the surface and to all the minerals underneath the parcel in question, in addition to the air space above it. Air rights are not owned to infinity, but the surface owner has the right to as much as he needs in connection with the use of his land.

A fee simple interest may be acquired by patent from a governmental authority, by warranty deed, by court order, or by devise or descent. The owner has the right to use and to occupy; the right of quiet enjoyment; and the rights to sell, mortgage, devise, lease, develop, improve, and subdivide. The fee simple owner may sever his land vertically or horizontally. One who sells only certain mineral strata, or to a limited depth, retains the right of access to the lower strata and to other minerals that he has not divested. By the act of severance, the surface and the minerals each become a distinct freehold estate subject to the laws of descent, devise, and conveyance. Multiple owners in undivided interest are not necessarily tenants in common, unless the law or an agreement so provides that status.

In most states it is a fact that purchasers are not bound by prior but *unrecorded* severances of surface or minerals unless the purchaser has actual knowledge of such a sale.[3] Knowledge of prior sale constitutes notice, even if the instrument has not been recorded.

Surface Ownership

Surface ownership from which the minerals have been severed carries the same full use and enjoyment rights that belong to the fee simple estate. Mere surface possession, however, is generally subject to the rights of the mineral owner to enter the premises for the purpose of extracting his minerals. It is not necessary that the surface owner relinquish the entire use of his surface for the mineral owner, but he may be required to allow as much use of the surface as is necessary for ingress, egress, and development. Hunt Oil Company v. Kerbaugh, 283 N.W.2d 131, illustrates that a mineral lessee who holds a dominant estate must show due regard for the rights of the surface owner. The mineral lessee has the right to reasonable use of that amount of the surface necessary for operations; but as a rule, the burden of showing any use to be unreasonable lies with the surface owner.

Minerals

A mineral owner—in fraction or in entirety—has the right to drill, explore, develop, and produce his minerals. Mineral ownership may be in undivided interests that have been conveyed, devised, or reserved under a deed or grant. One owner may not negate the right of development to another. Owners of fractional interests in minerals under one tract might be considered cotenants, depending on state statute. Each fractional owner may elect to lease or may pay his proportionate share of development costs as a working interest owner.

Mineral rights are executory; that is, the mineral owner may grant, sell, or convey in part or in whole. He alone may execute an oil and gas lease covering his interest. The mineral owner may sell all or part of his royalty (interest in production) and retain his executory rights. As a rule, minerals in place are considered real property; they are subject to real property laws and are transferable by the same formalities.

Most states allow the mineral owner who is not also the surface owner the right of ingress and egress to drill, mine, or develop his minerals.

While the majority of states do not tax nonproducing mineral interests, it is also true that in most states minerals are not lost by nonuse or abandonment or by failure to pay taxes that may be assessed against them. West Virginia, Pennsylvania, and Louisiana are three exceptions.

Louisiana is the only state in which laws of prescription govern mineral and royalty servitudes (severed interests). Prescription is the concept of reversion to the then present surface owner after a 10-year period of nonuse. The Louisiana Mineral Code provides that the owner of a mineral servitude under lands belonging to another party has a reasonable right to use as much of the surface as is necessary to explore for and produce his minerals in order to possess them.

West Virginia statutes allow severed minerals, previously untaxed or unassessed, to pass with a tax sale of the surface. Pennsylvania law distinguishes between "seated" (occupied) and "unseated" (unoccupied or wilderness) lands. Severed mineral interests under unseated lands may pass by a tax sale to the new purchaser of the surface, if the severed mineral interest had never been assessed.[4]

Minerals from which all interest in production (royalty) has been conveyed to others are commonly referred to as "dry" minerals. The dry mineral owner continues to own executory rights to lease and to receive bonus and rental monies, but he is divested of any interest in proceeds from production.

In the event that all minerals are removed, the mineral estate terminates. The space underground that was formerly occupied by the minerals reverts to the surface owner.[5]

Royalty

Royalty is defined as a share of the product or proceeds from the sale of minerals that have been developed or extracted (Simpson v. Burris, 365 P.2d 134). The word "royalty" literally derives from the king's royalty of one-fifth, which the King of Spain retained out of all production from the mines in the New World.

Royalty is generally considered to be an interest in real property. Royalty ownership encompasses the right of sale, devise, partition into undivided or divided fractions, descent, or mortgage.

By its nature, royalty is nonparticipating. This simply means that the royalty owner does not have an executory right to lease and he does not receive rentals or bonus monies due under the terms of an oil and gas lease. He may be entitled to a proportionate share of the shutin royalty paid under the terms of an oil and gas lease.

The royalty owner does not bear any of the costs of development or production. His is a passive interest that earns a proportionate share of proceeds from the sale of production.

Royalty ownership is created by grant or by reservation of the interest within a deed.

Palmer v. Crews, 35 So.2d 430, Miss., affirms that royalty in itself does not include a perpetual interest of oil and gas in the ground. A reservation of royalty is not a reservation of fee ownership. Royalty is a distinct estate. It presupposes a lease for its value. The royalty owner is dependent upon the fair practices and decisions to lease made by the owner of the minerals. Royalty interest may be conveyed for a term of years (and as long thereafter as production is obtained, if such wording is desired), or it may be conveyed "forever"—in perpetuity to heirs and assigns. An exception occurs in the state of Kansas, where perpetual royalty has been held to violate the rule against perpetuities.[6]

A mineral or fee owner may sell his royalty or deed it as a gift, fractionally or in its entirety. He may also sell his minerals and/or surface and reserve a royalty interest unto himself and his heirs.

Nonparticipating royalty interests, created as separate and severed entities, are paid out of the proceeds due to the lessor under the terms of the oil and gas lease that are in effect. Should the mineral owner elect to develop his own land without a lease, the nonparticipating royalty owner receives his proportionate fractional share of proceeds from production. Federal Land Bank of Houston v. U.S., 168 F.Supp. 788, defines a nonparticipating royalty as an interest in the gross production of oil, gas, and other minerals that is carved out of the mineral fee estate as a free royalty but does not have the right to participate in the execution of, bonus payable for, or delay rentals accruing under an oil and gas and mineral lease. The royalty owner holds no power to lease or to use the surface. The mineral owner is under an implied covenant to use fair dealings and diligence in his lease agreements, upon which the royalty owner depends. The duty of the holder of executive rights is discussed in Lamp v. Locke, 89 W.Va. 138, 108 S.E. 889 (1921), in which the plaintiff (royalty owner) based his claim against the mineral owner for failure to protect the acreage from drainage. The court held for the plaintiff, who had sued for fraud and asked for relief to require drilling upon his own land.

In a circumstance in which the lessee might be considered in breach of the implied covenant to develop, the royalty owner may or may not have the standing to sue the lessee. When he does not, the question arises whether it is the obligation of the mineral lessor to see that implied covenants are enforced. Williams and Meyers suggest that the better avenue is the law that allows the royalty owner to bring his own suit. "A Federal District Court in Louisiana held that the royalty owner cannot enforce implied covenants against lessee. . . .A West Virginia case has held that the royalty owner can require the lessee to drill a protection well."[7]

Working Interest

The working interest owner, or lessee, stands the full costs of exploration and development (less ad valorem taxes that may be attributable to the royalty interests). Net revenue interest (NRI) to the working interest owner is calculated by deducting the lessor royalty, any overriding royalty, and other burdens created by written contract from 100%. The "other burdens" might include production payments; net profit-sharing agreements; and carried working interests to a specified point in operations or to payout of the well possibly convertible to a full working interest or "back-in." Net revenue interest is of vital economic importance to the working interest parties in determining whether to drill and in calculating the rate of return on investment.

Working interest owners carry the liabilities applicable to their operations. These may include:

1. Injury to livestock or to surface features such as crops, structures, fences, trees, or grasses.
2. The escape of harmful substances from well operations.
3. Damage due to geophysical explosives.
4. Breach of contract under express or implied covenants.
5. Wrongful or negligent conduct.

Essentially, the working interest is an economic interest created by the execution of an oil, gas, and/or minerals lease. The mineral fee owner may develop his own minerals if he is knowledgeable in this area of the industry. Costs are high and risks are great.

The rights of the lessee under the terms of the lease may be sold, assigned, devised, or mortgaged as an interest in real property, or they may pass to heirs by descent. Specific rights and obligations of the lessee in the course of exploration and development are more fully discussed in Chapter 5.

Overriding Royalties and Carried Interests

Overrides may be created by conveyance or reservation out of an existing specified oil and gas lease. The overriding royalty is an interest in gross production. It carries no executory rights or rights to bonus or rental payments under an oil and gas lease affecting the lands covered. The overriding royalty owner has essentially the same rights as the nonparticipating royalty owner. He may sell, devise, sever into undivided fractional interests, or mortgage. As an interest in real property, an overriding royalty may pass by descent.

The override actually "overrides" the lessor royalty and is deductible from the net revenue interest over and above the royalty attributable to the lessor. It is liable for none of the costs of development. It expires upon termination of the lease under which the override is due.

A "carried" interest is created by written agreement with a lessee or working interest party. The participation is subject to its proportionate share of well costs and expenses and is an interest in net proceeds. The owner carries no personal liability for development costs as does a full working interest owner. Carried interests frequently revert to full working interest status at casing point or other designated time, such as payout. While "carried," costs are payable from production.

Overriding royalties and carried interests may be carved out of an entire eight-eighths or may be proportionately reducible to the fractional interest owned by the grantor.

Riparian Rights/Accretion and Reliction

Riparian, from the Latin *ripa,* meaning riverbank, refers to the rights of the landowner whose property borders on the bank of a stream or river. If the stream is not navigable, the adjoining landowner owns to the center of the riverbed, which changes with the course of the river through the years. In states where water is scarce, riparian rights are limited or nonexistent. They may be replaced by the "right of prior apportionment," giving legal use of the water to the first person taking it rather than to the owner of the land across which the stream flows. The property right, however, continues to the center of the stream or creekbed.

A reliction, or gradual recession of waters, leaves uncovered lands in its stead, adding to the gross acres of the adjacent landowner. This applies to minerals as well as surface. By the same token, accretion is the product of enlargement by the deposit of soil by moving waters. Eroding waters can likewise carry away a portion of the surface attributed to a landowner. Mineral ownership is surveyed by surface acres to the high-water mark or center stream.

Navigable Waters

Navigable riverbeds are generally owned by the state in which the river is situated, under specific grants from the United States government. Land abutting a river carries the right to the flow and reasonable use of the water. The landowner owns as property only to the usual high-water mark on his land. One of the many cases involving the bed of a navigable stream below the natural high-water mark is Halmadge v. Village of Riggins, 303 P.2d 244.

Adverse Possession

Title to land may be acquired against or without the consent of the record owner by continuous use and possession of an adverse owner. To establish adverse possession, one must show that possession was hostile, under claim of right or color of title, actual, open, notorious, exclusive, continuous, and for the statutory period of time, Fadem v. Kimbell, 612 P.2d 287 (Ct. App. Okla. 1979). The statutory period begins when the adverse possessor moves onto the land. The time set by law to establish title by limitation varies from state to state, from 10 to 50 years. Certain conditions must be met:

1. Possession must be adverse as opposed to permissive.
2. Possession must be open and notorious.
3. Possession must be continuous without interruption.
4. Possession must be for a period of time to establish ownership by prescription or limitation.

Payment of taxes by the record owner during the entire period does not necessarily preclude an adverse possessor's obtaining title.

"Constructive" adverse possession occurs when the adverse possessor holds under an instrument that he believes is effective but in actuality is not valid. Thus he holds by "color of title" or instrument of record that is faulty.

One period of adverse possession can be tacked onto the period of a successive possessor when passed from one to another by deed, will, descent, contract, or written or oral permission.

Adverse possession of the surface will not mature title to minerals that were severed prior to the entrance of the adverse possessor: Thornock v. Cook, 604 P.2d 934 (Utah 1979). In other words, prior severance of minerals precludes their adverse possession along with the surface. A reservation or severance of a fraction of the minerals severs only that portion. The remainder can be adversely possessed along with the surface. "Any severance of minerals or execution of an oil and gas lease *after* the beginning of the period of adverse possession would not, absent some ouster, interrupt the adverse possession."[8]

The basic reasoning protects the severed mineral owner in his lack of any legally protected interest in the surface other than his rights of ingress and egress. Adverse possession of severed minerals may be acquired only by the actual drilling, mining, or taking of such minerals.[9]

Life Tenants and Remaindermen

A life tenancy is considered an inferior estate of property ownership. It is generally considered that a life tenant may use, occupy, possess, and enjoy the premises during his life. He may not sell, lease, or in any way alienate the property without the joinder of all remaindermen. The corpus of the life estate must be preserved, and the life tenant may not commit waste.

Generally, the life tenant does not have the right to produce oil and gas unless production was obtained prior to his position. Under the "open mine" theory, he may receive such income as was in effect at the inception of his status.[10] Otherwise, the life tenant may expect only the use of the land and perhaps the interest accrued from proceeds received during his life estate.

A joint lease, executed by life tenant and remaindermen, necessitates a delay rental agreement unless the parties agree that the payments shall be made to them jointly at the designated depository bank. The remaindermen are usually entitled to bonus and royalty unless these monies are placed in suspense to accrue interest for the life estate.[11]

"A remainder is vested when there is a person in being who would have an immediate right to the possession on the termination of the intermediate estate. It invests him with a present interest and a right to future enjoyment," Capels v. Ward, 107 Tex. 341.

A remainderman must join in or ratify any deed, lease, or other conveyance of his interest that might be executed by the party holding a life estate in the property. During the life estate, the remainderman may enter his property to collect rents, to determine if waste is being committed, or to make necessary repairs.

A life tenant may assign his life estate and create in his assignee a like life estate that would terminate on the death of the assignor. He is obligated to pay assessed taxes and interest on any mortgage due. If a life estate is created in several grantees, the life estate exists until the death of the last survivor. A life estate terminates on the death of the life tenant.

Joint Tenancy and Tenancy in Common

The three usual classifications of concurrent ownership in real property are joint tenancy, tenants in common, and tenancy by the entirety.

Joint Tenancy. In joint tenancy two or more equal owners of the whole exist. Joint tenancy is created by grant or by will; it cannot be created by descent. Upon the death of one owner, the survivor(s) owns the entirety. This is known as joint tenancy with right of survivorship, and is commonly referred to as JTROS. JTROS is not recognized in states with community property laws. Each joint tenant has the right to dispose of his fractional interest. The joint tenancy may be terminated by a joint conveyance or by partition.

Tenants in Common. Tenants in common are two or more persons who jointly own property concurrently. The undivided interest of each may be different, and each may sell or devise his interest. Most states allow one tenant to lease and develop without the joinder of the other undivided owner(s). However, Louisiana, West Virginia, and Illinois have laws which provide that all must join in any lease on the jointly owned premises in order for development to take place.[12] The estate may be destroyed by partition or by all tenants jointly conveying to one party. Tenancy in common does not carry the right of survivorship, but each interest is inherited by the heirs or devisees of the individual owners.

Tenancy by the Entirety. Tenancy by the entirety exists only between a husband and wife. Neither spouse may dispose of his or her interest without the joinder of the other. Survivorsip exists. If divorce terminates the estate, the parties become tenants in common.

COMMON LAW CONCURRENT TENANCIES COMPARED *

Kind of tenancy	How created	Typical words in deed or will	Unities present	Interest owned by tenant	Power of disposition	How can disposition be made	Rights on intestacy	How destroyed
Joint tenancy	By act of the parties, deed or will	A to "B & C and their heirs as joint tenants with the right of survivorship and not as tenants in common."	Time Title Interest Possession	All tenants as a unit own the whole, *joint* ownership	All may join and dispose of whole or each tenant can dispose of share he did not own as such	By deed only and not by will. Survivorship defeats effect of will.	If only one survivor he continues to own but in severalty. If more than one survivor they continue to own in joint tenancy.	1-One tenant conveys his interest. 2-By partition in kind among tenants. 3-By any act which breaks any unity.
Tenancy in common	By act of the parties, deed, will, or by law	A "to B and C and their heirs share & share alike as co-tenants."	Possession	Each tenant owns an undivided portion which portions are not necessarily equal	Each tenant can dispose of his undivided share or part thereof	By deed or by will	Heir or heirs inherits undivided interest of deceased co-owner	1-By partition among tenants 2-By uniting all titles in one tenant in severalty by purchase or otherwise
Tenancy by the entirety	By act of the parties, *deed or will*	A "to H & W and their heirs."	Time Title Interest Possession Person	Husband and wife as a unit own the whole, *joint* ownership	Both husband and wife must join in conveyance	By deed only and not by will. Survivorship defeats effect of will.	Survivor continues to own all but in severalty	Divorce terminates the tenancy and makes them tenants in common
Tenancy in coparcenary	By law of inheritance	A dies intestate leaving 3 daughters, B, C & D, his only heirs	Title Interest Possession	Each tenant owns an *undivided* portion which portions are not necessarily equal	Each tenant can dispose of his undivided share or part thereof	By deed or by will	Heir or heirs inherits undivided interest of deceased parcener.	1-By partition 2-By conveyance by one parcener 3-By whole descending or vesting in one parcener

* Of course such an outline at best is not complete but is intended merely to set forth and compare the outstanding characteristics of the tenancies mentioned. Many variations exist among the jurisdictions. The chart reflects the characteristics prevailing in the majority of jurisdictions that recognize the particular estate.

Fig. 2–1 Common law concurrent tenancies compared (Reprinted from Myron G. Hill, Jr., Howard M. Rosen, and Wilton S. Sogg, *Smith's Review, Legal Gem Series*, 3d. ed. (St. Paul, Minn.: West Publishing Co., 1978), p. 116, with permission of the West Publishing Company.)

Curtesy and Dower

Curtesy is a life estate created for a husband, whereas dower is a life estate created for a wife. These estates do not exist in states that have community property laws. Under English common law, a husband is entitled to all lands owned by himself and his wife. Curtesy exists provided they have issue born alive who will inherit the lands.[13] Curtesy grants to the husband a one-third life estate in their properties on the death of his wife.

The widow's dower is a one-third life estate in lands owned by the husband during their marriage. This right cannot be destroyed by the husband by way of conveyances that he might choose to make during his lifetime unless the wife joins in the conveyances or releases her dower voluntarily. Individual state laws should be examined to determine if these estates survive or have been modified.

Easements and Equitable Servitudes

Easements and rights-of-way are considered inferior estates of ownership in land. Easements may be for use or for profit or both. The extent of the rights is determined by the terms of the conveyance granting or reserving the easement or right-of-way. Mineral owners have an implied easement of ingress and egress to the surface estate for the purpose of exploration and development.

Easements may be created by grant or by terms of reservation if a landowner sells his property and reserves an easement for a definite purpose. Easements may be for a term or in perpetuity and may be exclusive or nonexclusive. They may be acquired by adverse possession for the period necessary to prescribe. An easement may terminate by conveyance, or abandonment, or use that is inconsistent with the terms and provisions of the easement grant.

Land use restrictions that are enforceable by law, such as deed restrictions for subdivisions and restrictions created by zoning laws, are referred to as equitable servitudes. Servitudes originating from eminent domain are created by condemnation of private lands for public use. The exercise of eminent domain by governmental authority must be reasonable, and the landowner must be justly compensated for the taking under the Fifth Amendment of the Constitution of the United States: "nor shall private property be taken for common use without just compensation."

If lands are dedicated by private owners for public use, the instrument of dedication should be examined to determine the rights conveyed, whether the dedication is terminable and on what conditions, and which rights (such as minerals) are reserved to the grantor.

The Founding Fathers of the United States placed great emphasis on the connection between liberty and private ownership of property. One of the major goals in the drafting of the Constitution was to protect the right of the individual to use and enjoy his property. In addition, the Fourteenth Amendment prohibits that any state government may deny a person the use of life, liberty, and property without due process of law.[14]

Indian Lands

During the period of colonialism, the majority of European countries gave sovereign title to lands to the discoverers.[15] Subsequently, the right and title to Indian lands in the continental United States came to this country subject only to rights of occupancy that were granted to certain tribes. The Bureau of Indian Affairs, established in 1824 as a part of the War Department, became a division of the Department of the Interior in 1849. The Indian Removal Act of 1830 gave the president power to exchange lands west of the Mississippi River for Indian-occupied lands situated within the boundaries of any state and acted to segregate the Indian tribes.[16]

Various pieces of legislation affected the titles to Indian lands in the United States. Reforms brought a more cohesive policy with the Indian Reorganization Act of 1834 and the General Allotment Act of March 8, 1871. A fiduciary relationship existed between the federal government and the Indian due to the requirement that government consent was necessary prior to the Indian's transfer of his land.[17] Three methods of placing ownership in the Indian were created: by treaty (prior to 1870); by executive order; and by act of Congress (since 1919), 41 Stat. 3,34 (1919). A ruling in Spaulding v. Chandler, 160 U.S. 394 (1895), stated that the manner in which a reservation is created is of no consequence in determining title in tribes.

The Wheeler-Howard Reorganization Act of 1934 provides for limited self-government by tribes and for the secretary of the interior to issue charters of incorporation. Tribal councils may direct activities dealing with local, state, and federal government. This act includes the power to sell, mortgage, or lease reservation lands for terms limited to 10 years (25 USC 477, 1976).

The two types of Indian lands are tribal lands and alloted lands. Tribal lands are those originally vested in the Indian tribe and are called aboriginal lands. Title to these lands was sustained in the tribe under Cramer v. United States, 261 U.S. 219 (1923), even though Congress had granted the land to a railroad. "Where aboriginal title is not constitutionally protected from divestiture, the more common treaty-based title receives the full protection of the Fifth Amendment,"[18] causing treaty lands to border on fee simple absolute title.[19]

In Crow Nation v. United States, 81 Ct.Cl. 238 (1935), the court held that the granting of lands by executive order to Indians gave only the right to reside on the lands. Indian lands acquired by other means are generally considered subject to protections under the Indian Reorganization Act of 1934.[20]

The General Allotment Act of 1887 provides for assignment of lands to individual Indians. The lands must be limited, comprising no more than 40 acres of irrigable land, 80 acres of agricultural land, or 160 acres for grazing purposes. The individual obtains his patent, and the title is held in trust by the United States government for 25 years. (The trust may be extended by statute or by executive order.[21]) However, the United States continues to retain legal ownership of the alloted land. The Indian holds only equitable title until all restrictions are lifted (Oklahoma v. Texas, 258 U.S. 574, 1922). Upon expiration of the trust period, the secretary of the interior may, at his discretion, issue a patent for fee simple title to the individual Indian for his alloted land.

Exempt from the General Allotment Act were five so-called civilized tribes: Cherokees, Creeks, Choctaws, Chicasaws, and Seminoles, as well as the Osage and Senecas.[22] These particular tribes were alloted lands by patents with 25-year restrictions (34 Stat. 539, 1958).

Certain treaties and statutes specifically provide that the minerals will not vest in the tribe. Other grants provide for "absolute use and occupation," a phrase construed to include interest in minerals and timber under United States v. Shoshone Tribe, 304 U.S. 111 (1938).[23] The general rule pertaining to alloted lands is that it does include mineral and timber rights, subject to certain restrictions.

"Care should be taken to examine all relevant treaties, statutes, executive orders, and amendments" before proper determination can be made of mineral ownership in Indian lands.[24]

Public Lands

Public lands in this concept include federal and state domestic lands open for sale, use, or other disposition. They do not include all lands under public domain.

Governmental grant is the true source of title in the United States. A certificate issued by the federal government indicates that the receiver is entitled to a patent. A patent differs from a deed in that delivery is not essential to pass title. Acceptance by the grantee is presumed.[25] A patent may be issued to the entryman or to his heirs or assigns. A patent is the highest evidence of title. Mutilation or destruction of the instrument of patent does not impair its validity.[26]

In the United States territories, discovery on American soil, followed by possession, gave title to the government taking possession against all other claimants, including Indian tribes. Titles into the United States were also obtained through treaties and cessions from foreign governments. Private ownership rights were not invalidated with changes in sovereignty.

The federal Constitution empowered Congress to dispose of, sell, or regulate the public domain without limitation.[27] The original 13 states came into ownership of their public lands after the Revolutionary War through the treaty with England. Certain of those public lands were granted to the federal government.[28] Others were retained by the colonial states. Titles to public lands in other states were acquired by the federal government through various enabling acts.

Laws governing public lands are statutory and subject to frequent change. Today public lands are administered by the Bureau of Land Management (BLM), a division of the Department of the Interior, in concert with the Minerals Management Service. The BLM has the power to classify, sell, withdraw, lease, or use for the public good and benefit any of the lands in its domain. Historically, the federal government has retained all mineral rights under public lands owned or acquired, with the exception of lands patented to individuals, to railroads, and to the states (including water bottoms of navigable rivers).

Public lands currently administered by the federal government comprise more than 470 million acres and are principally situated in the western United States, including Alaska. The lands are classified according to "best use" and are held in trust for the public interest. Some 525,000 acres are maintained for parks and national monuments.

The Mineral Leasing Act of 1920 provides that all lands within known geologic structures (KGS) or in producing fields are to be leased on a competitive basis. All other lands are leased noncompetitively by simultaneous bidding or filing. The Federal Land Policy and Management Act of 1976 gives the Bureau of Land Management its statutory base. Additionally, it requires the secretary of the interior to review for possible withdrawal (from leasing) by 1991 many thousands of so-called wilderness lands in 11 western states. These wilderness areas are of particular concern to the oil industry. Certain BLM leases already contain wilderness protection stipulations.

Control and disposition of state lands are determined by individual state laws. The power to govern designated sectors is frequently vested in officers, commissioners, and boards.

A patent granted by a state to an individual is evidence of legal title. It is effective on the date of issue.[29] The patent carries a seal and the signature of the governor, secretary of state, or officer charged with executing land grants. A state patent generally contains no warranty provision.

Statutory provisions were made by Congress for school land grants to new states on their admission to the Union. Under the Morrill Act of 1862, Congress granted each state 30,000 acres for each representative that the state had in Congress, for the purpose of establishing land-grant universities. The land was intended to be sold and the proceeds used to build and maintain agricultural and mechanical colleges. Some states sold their lands; others retained the largest part. Minerals to thousands of acres were thus passed from the federal to state governments. Other lands passed to the states under "selection" systems. The leasing of school and other state lands is authorized by statutory procedure.

References

[1]Howard R. Williams and Charles J. Meyers, *Oil and Gas Law* (New York: Matthew Bender, 1975), pars. 203.1–204, pp. 16–17.

[2]J.F. Womack, "Evolution of Land Law," *Advanced Land Practices Institute, Houston* (Ft. Worth: AAPL Educational Programs, 1982).

[3]*54 American Jurisprudence, 2d, ser.* (San Francisco: The Lawyers Co-Operative Publishing Co., 1980), pars. 102 ff.

[4]Edwin Stitt, "Eastern Overthrust Belt," *Landman* (June 1982), pp. 29–30.

[5]*54 American Jurisprudence,* pars. 134 ff.

[6]Williams and Meyers, par. 303.6, p. 75.

[7]Ibid., par. 339.3, p. 189.

[8]*54 American Jurisprudence,* pars. 134 ff.

[9]Williams and Meyers, par. 224.1, p. 51.

[10]Earl A. Brown, *The Law of Oil and Gas Leases* (New York: Matthew Bender, 1958), par. 208, p. 12.

[11]Richard W. Hemingway, *The Law of Oil and Gas* (St. Paul, Minn.: West Publishing Co., 1971), p. 178.

[12]American Association of Petroleum Landmen, "The Oil, Gas & Mineral Lease," *Advanced Land Practices Institute, Houston* (Ft. Worth: AAPL Educational Programs, 1982).

[13]Chester Smith, *Legal Gems* (St. Paul, Minn.: West Publishing Co., 1972), pp. 151–162.

[14]James MacGregor Burns and Jack Walter Peltason, *Government by the People* (Englewood Cliffs, N.J.: Prentice–Hall, 1972), pp. 180–183.

[15]Johnson v. McIntosh, 21 U.S. (8 Wheat.) 503 (1823).

[16]Act of May 28, 1830, ch. 148, 4 Stat. 411, sec. 3.

[17]The Indian Non–Intercourse Act of 1834; and United States v. Southern P. Trans. Co., 543 F. 2d 676.

[18]Gary Boyle, "Indian Lands," *Landman* (October 1981), p. 11.

[19]Ibid., p. 11.

[20]Ibid., pp. 11–12.

[21]Ibid., p. 12.

[22]Ibid., pp. 12–13.

[23]Ibid., p. 13.

[24]Ibid., p. 14.

[25]U.S. Constitution, art. IV, sec. 3.

[26]State v. McAlhaney, 220 NC 387, 17 S.E. 2d 352.

[27]McBride v. Schurz, 102 U.S. 378.

[28]Wright v. Roseberry, 121 U.S. 488; Bickness v. Comstock, 113 U.S. 149.

[29]Ex parte Kuhtman, 24 S.C. Eq 257.

CHAPTER THREE

Instruments of Conveyance

A deed is a conveyance that intends to pass title to a freehold tenure in land, historically called "seisin," meaning the ownership of land with attendant rights of possession and conveyance.[1] Early common law required the ceremonial delivery of a conveyance, by "livery of seisin," in the presence of witnesses. The writing was duly executed and delivered, complete with seals affixed. It was not the writing but the delivery that transferred the title. The writing was merely the evidence of the transfer.[2] The common law did not require witnesses to validate a deed. Some states do now require an attestation, and others require both attestation and acknowledgment.[3] The acknowledgment is necessary to qualify the instrument for recordation in the county and state where the property is situated.

Conveyancing in the United States is accomplished by an instrument "generally drawn with covenants of title, and which accomplishes transfer of title to real property by a deed expressing the intent of the owner to convey it."[4] "Execution" of a deed constitutes the signing, sealing, attesting, acknowledgment, and voluntary delivery of the instrument. Delivery is necessary to pass title.[5] An undelivered deed does not divest the grantor.[6]

Acceptance by the grantee is vital to the transaction. Even a gift deed that imposes no obligation on the grantee requires his acceptance.[7] If it is against his wishes or his best interest to accept, the grantee may file a disclaimer that reinvests the interest back to the grantor.[8]

An effective deed depends upon the legal right of the grantor to sign and convey; this is the grantor's "executory" right. The mineral owner alone or his legal representative is empowered to execute an oil and gas lease. A party cannot convey what he does not own or more than he owns. Therefore, a conveyance of three-fourths of the undivided minerals from a grantor who owns only one-half would pass title only to the one-half interest.

A deed may be signed in one's own handwriting, or under special circumstances another may sign for him. An illiterate may make his cross or mark if it is properly witnessed and acknowledged.[9] The signature does not have to be in ink, but it certainly should be for protection and permanency. A signatory party does not have to sign his full name for the instrument to be valid, as long as there is no doubt as to his identity.[10] However, the majority of legal advisors require curative measures to establish the correct name of the individual or entity. Proper signature, conforming to the record name of the owner, saves confusion and doubt.

An owner may authorize another to act on his behalf. A formal authorization is required in most states and is known as a power of attorney. This authorization designates one party to act as the agent and attorney-in-fact for another party, for general or designated purposes. A power of attorney is sufficient only for the intentions and powers specifically granted. Trustees, administrators, and guardians are empowered to sign on behalf of other parties.

Corporations are represented by the signatures of certain officers who are authorized by corporate resolution of the board of directors. A corporate resolution is rarely placed of record. In any deed that states "unto the said grantee, his heirs and assigns," should the grantor be a corporation, company, or other entity of business, it would have no heirs. The word "successors" would be substituted for "heirs."

If several persons own undivided interests in the same lands, execution by one or more does not convey the interest of the unsigned parties. Deeds may be executed in multiple counterparts.

Recordation serves as effective delivery when the grantor has the instrument properly recorded, an indication of his full intent to convey title.[11] Upon recordation, the grant becomes effective. In the state of Kansas, a mineral deed must be recorded within 90 days of the date of execution or within 90 days from the date the property is registered for tax purposes. This law is not applicable to royalty interests.

Prior knowledge of the existence of an unrecorded instrument constitutes actual notice thereof.[12] A subsequent grantee is bound by the terms of any such instrument of which he has knowledge or information. It is not necessary that federal and state leases be placed of record in the county or parish where the lands are situated, but local recordation assures good and effective notice of the transaction and serves to maintain a complete chain of title in the local records.

Deeds are comprised of formal parts, but all are not legal requirements for a transfer to be valid.[13] In addition to the grantor and grantee and the description of the property to be conveyed, there is a granting clause with apt words of grant to express the intentions and motives of the grantor.[14]. The tendency of the courts today is to relax the formal rules and to accept any wording that clearly states the agreement between the parties.[15]

Following the names of the parties, a consideration is expressed. It may be a "good" consideration, founded on love and affection, moral obligation or duty, i.e., a personal consideration.[16] Most often the consideration is "valuable," by some nature of payment. Many deeds cite "ten dollars" or "one dollar" and O.V.C., meaning "other valuable considerations," in lieu of the actual amount of the transaction in order to prevent the public disclosure of a private negotiation of terms. Implied and express convenants serve as the "other valuable considerations."

Proper legal description may be by lot, street, house number, city, county, state; by adjoining land owners; by metes and bounds; township and range; reference to map or plat records or to deed records as long as the description is sufficient to accurately describe and locate the lands. A deed may refer to another deed for its terms as long as the deed referenced sets out a proper description and information.[17] For instance, a deed reference for a metes and bounds description should correctly refer to an instrument containing a full metes and bounds description, not to another deed reference.

Indefinite words are often found in property descriptions. "Adjoining" implies a contiguous border, even if it is only a common corner. An adjoining boundary does not have to coterminate. The word "adjacent" implies contiguity; the phrase "in the vicinity of" does not necessarily mean that the lands abut.

Descriptive phrases and words of precaution are common in order to cover unintentional and unimportant inaccuracies in the cited description.[18] Certain of these phrases and words are "more or less," "on or about," and "estimated." The words "more or less" are purported to have first been accepted by a court in the California territory when ropes were used to measure distances. When a rope was wet, it would stretch farther than it would when it was dry.

> If it happened that the lariat was drawn through wet grass, it would lengthen under the strain as they proceeded and the grantee would

> then secure generous boundaries, and an enlarged area for the square leagues mentioned in the grant. This excess of quantity was, however, sustained by the courts upon the qualifying phrase in the grant, "more or less."[19]

The words "part of" or a "divided fractional part" imply an intention to convey in severalty or partition. "Undivided" indicates a contenancy under the whole.[20]

Following the description is the habendum clause. In the habendum clause the estate granted under the granting clause is often enlarged or qualified. This clause acts as a control on the granting clause and is considered to be the last expression of the intent of the grantor.[21] The habendum clause in an oil and gas lease specifies the term of grant as it does in a deed. The "thereafter" provision has been construed as a special limitation to the term.[22]

After the habendum, the words of inheritance "his heirs and assigns forever" are not required in all states. In situations where such wording is required and is omitted, a reservation of interest could be construed to be for the life of the grantor only.[23]

"Exceptions" set out from the granting clause or in the description should be strictly defined. Each exception constitutes a withdrawal of part of the grant. Exceptions cited within the warranty clause serve to protect the grantor from breach of warranty. Exceptions save for the grantor an interest or right already in existence. However, in Oklahoma "except" clauses in the habendum do not act to reserve an interest to the grantor—they simply act to exclude that interest from warranty. Kansas and California courts have reacted similarly to provisions of exception. Legal counsel is strongly recommended in the preparation of conveyances.

A "reservation" creates a distinct and independent ownership of separate or severed interest that did not exist prior to the conveyance in which it is described. Reservations may be of mineral or royalty interest, overriding royalty, water rights, a life estate, an easement, or rights to timber and growing crops.

A rule of common law is that interests may not be created to a stranger to the title by means of exceptions and reservations in a conveyance. The rule might be set aside if the stranger acted as a party to the conveyance by executing the instrument. Decisions are divided as to whom the reserved or excepted interest would rightfully belong.[24]

The state of Pennsylvania does not consider oil and gas to be minerals. Therefore, a reservation of minerals in a deed conveying

Pennsylvania lands does not reserve oil and gas unless these products are specifically mentioned.[25]

"Subject to" clauses describe encumbrances presently existing on the conveyed premises. The owner of less than 100% interest who conveys and retains a partial interest should convey "subject to" any prior reservations.

Williams and Meyers cite Hoffman V. Magnolia Petroleum Co., 273 S.W. 828 (Tex., 1925), as a decisive case in influencing "subject to" clauses. The court resolved a conflict between the granting clause and the "subject to" clause in a deed as two separate grants.[26] Louisiana has rejected the Hoffman case theory, but controversies exist over deeds with ambiguous and conflicting provisions.

The purposes of the "subject to" clause are to protect the grantor from breach of warranty, to lend precise wording to clarify the grant, and to clarify rights that may only pass by express wording.[27]

Warranty Deed

Warranty and other covenants of title guarantee that the grantor owns merchantable title with the right to convey. A warranty deed contains a full warranty of title. A special warranty deed contains words of limitation, such as providing for warranty of title by, through, and under the grantor only. This restricts the grantor's liability to his own acquisition of the lands and interest conveyed. A warranty deed (or special warranty deed) conveys all of the right, title, and interest of the grantor, subject to reservations set out, in and to the surface or surface and minerals. Silence as to a reservation of minerals will convey the minerals owned by the grantor. A conveyance of more interest than the grantor owns does not enlarge the title to the grantee. That is, one cannot convey more interest than he owns.

A deed closes with a testimonium: "In witness whereof"; an effective date if it has not been entered at the top of the deed; a signature line; and an attest, if necessary. Style of acknowledgment should conform to the requirements of the laws of the state in which the property is situated. Acknowledgments are prepared for individuals, for persons having the capacity to sign for another, and for corporations. Acknowledgment forms are discussed in detail in Chapter 5.

Unintentional errors or material mistakes may be corrected by the preparation, full execution, and filing of a correction deed or by a revocation placed of record if the grantor and grantee mutually agree to do so.

Quitclaim Deed

A quitclaim deed is a form of release effected by the words "remise, release, and forever quitclaim."[28] It is filed for the purpose of excluding the grantor from any future claim on the title as to his present interest. Unless an after-acquired or future interest is specified, it is not included. The grantor conveys only that interest he owns at law or in equity—not necessarily a fee—at the time that the quitclaim is made. Quitclaims may be filed to release a dower interest or a remainder interest or to remove a cloud on the title.

Easement Deed

An easement deed conveys a right or privilege of use. It does not usually convey a fee interest of the surface or the minerals. However, in the absence of specific language of limitation, an easement deed could possibly be construed to pass title in fee.[29]

Testamentary Grant

A testamentary instrument, or will, passes title to lands after the death of the maker. A will does not function to convey title while the testator (maker) of the will is still alive.[30]

A will that does not specify all properties passes unnamed properties under a "rest and residue" provision. In the absence of a "rest and residue" clause, properties not specifically devised pass title under the laws of descent and distribution effective in the state where the property is situated.

Mineral Deed

A mineral deed is a conveyance of interest in real property to transfer title of minerals in place.[31] The mineral deed may sever the mineral interest from the surface, creating a separate freehold estate. The deed may transfer all or a fractional portion of the grantor's undivided interest. The mineral deed may be forever unto the grantee, his heirs and assigns, or it may be terminable. A term deed usually provides that on a certain date if no production in paying quantities is being produced, the grant terminates. Commonly, term mineral deeds are for 15 or 20 years and "as long thereafter" as oil or gas is "found." The word "found" has been construed to mean "produced."[32] (This wording creates a determinable fee.) If the term mineral deed has no "thereafter" clause, an oil and gas lease taken from the term mineral owner should be ratified by the reversionary owner in the event that production extends beyond the mineral term. Temporary cessation of production that does not terminate the lease under which production is obtained may not terminate the mineral grant that is into its secondary or "thereafter" term. After cessation of production, future production cannot revive prior rights.[33]

A landowner who executes a deed with no reservations or limitations conveys the surface and the minerals below that have not previously been severed. To sever or reserve the minerals, express wording must be used.[34] Transfers or assignments of interest in minerals or royalty are subject to present encumbrances and all prior reservations of record.

When fractional interests are conveyed or reserved, the fraction may be calculated on the whole or on the fractional interest of the grantor at the time of the grant. (The decision in King v. First National Bank, 144 Tex. 583, 192 S.W. 2d 260, rejected a grantee's claim for the lesser fraction. The grantor had reserved an undivided one-eighth royalty under a conveyance of one-half mineral interest. The grantee had contended that only one-eighth × one-half royalty was thereby reserved.) The intent of the parties *should* be specified exactly, but alas, the intent is often unclear.

Confusion exists in the interpretation of the words "mineral" and "royalty." The terms are used loosely and interchangeably in many older conveyances of record. Courts have relied on the intent of the parties when it can be ascertained and have taken into consideration the familiarity of the parties with oil and gas terminology.

The interpretation of many mineral and royalty deeds of record is difficult, even for the experienced landman. What phraseology distinguishes the mineral conveyance from the conveyance of royalty only (a share in production)? Key language that has been most frequently construed to refer to minerals is "in and under and that may be produced," even though the deed may use the terms "mineral" and "royalty" interchangeably. Older deeds styled "royalty contracts" may intend to convey a mineral interest.

A clue is often found by searching the title behind the ambiguous instrument. What did the grantor own? That is, what interest was conveyed to him originally? Oklahoma courts have long recognized that the terms "mineral" and "royalty" have created ambiguities. It appears that Oklahoma law construes a grant or reservation of a fraction of royalties (e.g., one-fourth royalties) to mean minerals and a fraction of a specific royalty to mean royalty is conveyed (e.g., one-fourth of a one-eighth royalty).[35] Kansas court decisions have fluctuated in interpretations of the intent of the grantor. Generally, if a deed appears to convey the right to execute oil and gas leases, with attendant rights to receive bonus and rental monies, it is likely that the deed transfers an interest in the minerals.

The intent of a grantor to convey an undivided fractional interest in what he owns results in instruments that are real puzzlements. The grantor who owns an undivided one-half mineral interest may desire to sell one-half of his half, or one-quarter. A deed that stipulates a grant of an "undivided one-half interest" and then describes the entire premises would serve to divest the grantor of all he owned. "Fifty percent (50%) of my undivided one-half interest" is more specific, and would result in the sale of one quarter of the total.

Royalty Deed

The technical meaning of "royalty" is construed to be an interest in production, free of costs and dependent on the existence or anticipated existence of an oil and gas lease. If royalty is conveyed with no oil and gas lease in effect, and if the fractional "royalty" conveyed is not specifically defined, the royalty grantee might be asked to ratify any lease granted by the mineral owners. Language would be included in the ratification to exclude the royalty owner from an interest in money payments other than royalty from production.[36]

In the absence of a specific definition within the instrument, an examination of the entire wording or circumstances must often determine whether minerals or royalty were meant to be conveyed. If the reference designates a percentage of production, it might be concluded that royalty is conveyed.[37]

Instruments transferring "all" royalty have been construed to be minerals in place, as in Toothman v. Courtney, 62 W. Va. 167, 58 S.E. 915. Under Palmer v. Crews, 35 So.2d 430, a devise of "all my royalties" was found not to be minerals in place. Under Davis v. Hurst, 150 Kan. 130, the court held that a reservation of one-half of the royalties did not mean minerals and did not entitle the grantor to a share of the money payments.

The essential difference is that a royalty owner receives nothing under the terms of a lease until production is found. The mineral owner shares in rental and bonus monies and has the executory right to grant a lease. The royalty owner does not. An instrument conveying "royalty" that includes rights to the rentals or bonus to extend the life of the lease or that conveys a future interest in money payments has been held to transfer an interest in the minerals.[38]

As a rule, a deed that transfers "all of the oil, gas, and other minerals in and under and that might be produced" will cover minerals. A conveyance subject to an oil and gas lease which states that it grants an interest in royalty "produced and saved" would likely be a royalty deed.[39] In the granting clause a mineral grantor may sell a fraction of the oil "in and under and that might be produced" and in a later clause in the deed grant the same fraction times the royalty payable under an existing lease. The court held under Little v. Mountain View Dairies, 217 P.2d 416, that such language is not inconsistent and that the granting clause prevailed. The transfer of rights to lease or to receive monies are attributes of ownership that disclose an intent to convey oil and gas in place, i.e., minerals.[40]

Royalty conveyed by deed is out of and a part of any lessor royalty created under the terms of an oil and gas lease.

Lease and Assignment

The basic difference between a mineral deed and an oil and gas lease is that the deed is evidence of a sale, a grant of a separate estate. Certain states' statutes consider the oil and gas lease to be a "sale" of real property; but under a lease the lessee may take action to keep the lease in force. Production is immaterial to the effectiveness of the mineral deed. A lease has an implied covenant to develop; a deed does not. The conveyance of mineral interest includes the right to enter upon the premises for ingress and egress, to explore and produce, even without an express convenant to that effect.[41] A lease "demises and leases" for a specified period of time and allows the lessee reasonable use of the surface for installations. An instrument granting only a right to enter upon the property and extract the minerals might be interpreted to be a license.[42] Under surface deeds or leases, the surface is defined as that part of the land capable of being used for agriculture (notwithstanding any other use to which it is put).[43]

The lease instrument should be in writing and signed by the party against whom the rights are charged (lessor) or by his authorized agent. Unless the statute provides that oral contracts are valid, no oral agreement, lease, or contract is enforceable.[44] Some statutes require that a lease for a term of more than one year must be in writing.[45] Any extension of the terms should also be by written agreement.

The recording of an oil and gas lease provides constructive notice of the transaction. Assignments, transfers of interest in the lease, or mortgages should also be recorded because of their inherent nature as interests in real property and to protect a future purchaser.[46]

A nominal cash payment as consideration under an oil and gas lease has been found to be sufficient when joined by the promise of royalty payments out of production.[47] However, a one-dollar consideration has been judged to be insufficient when the drilling of a well is left to the discretion of the lessee, i.e., if it is a unilateral contract not executed by the lessee and contains no drilling commitment.[48]

Rights of a surface lessee or of a grantee under an easement deed must be taken into consideration by the owner of the leasehold estate as he prepares to enter the property for exploration and development. Surface waivers and damage releases are discussed in ensuing chapters.

Logue v. Von Almen, 40 N.E. 2d 73, illustrates several points of interest regarding oil and gas conveyances. It claims a voidable deed is good until set aside by the court. The voidable deed passes defeasible title (subject to being defeated by suit). Under Logue, if a deed indicates it is to be executed by several parties and only certain of the grantors sign, the deed passes title only to the interest signed. For example: If the deed purports to convey an undivided one-half interest, but all owners of the one-half do not sign, and perhaps only one party owning one-fourth undivided interest in the one-half executes the deed, then it conveys only one-fourth × one-half (the fractional part of what the grantor owns). Logue, v. Von Almen also discusses the fact that an acknowledgment taken before a notary public who has a financial interest in the property will be nullified.

Assignments or subleases include transfers of interests in real property in whole or in part. An assignment may be construed as a grant, but more frequently it is intended as a transfer of intangible rights. The assignor is the party transferring the right or interest; the assignee is the taker under the assignment.

Assignments may cover future or contingent interest or rights existing only in possibility, such as an assignment or overriding royalty executed prior to the discovery of production.[49] Assignments of oil, gas, and mineral leases transfer operating rights, leasehold working interest, and attendant net revenue interest. In certain jurisdictions, the assignment carries the transfer of responsibility and the liability to comply with the terms of the oil and gas lease to which the assignment is made. Other statutes, such as in Louisiana, provide that the original lessee or assignor remains liable in the event of failure of assignee to perform or comply.

Court Decrees/Miscellaneous

Title to property may pass by court decree under property settlements and partitions, divorce proceedings, boundary disputes, or other civil suit. A trespass to try title suit does not actually transfer interest, but serves to confirm interest as claimed by the plaintiff.

Title may pass by foreclosure of lien, tax sale, sheriff's deed under bankruptcy or by government taking under laws of eminent domain.

References

[1]Henry Campbell Black, *Black's Law Dictionary,* 4th ed. rev. (St Paul, Minn.: West Publishing Co., 1968), p. 1524.

[2]Green v. Liter, 8 Cranch (U.S.) 229.

[3]Byrd v. Bailey, 169 Ala. 452.

[4]*23 American Jurisprudence,* 2d ser. (San Francisco: The Lawyers Co-Operative Publishing Co., 1980), par. 15.

[5]Turlington v. Neighbors, 222 N.C. 694.

[6]Young v. Guilbeau, 3 Wall (U.S.) 636; Griswold v. Griswold, 42 So. 554.

[7]Talty v. Schoenholz, 154 N.E. 139.

[8]Halleck v. Halleck, 337 P.2d 330.

[9]Loyd v. Oates, 143 Ala. 231.

[10]Zann v. Haller, 71 Ind. 136.

[11]Tompkins v. Wheeler, 16 Pet (U.S.) 106.

[12]Leslie Moses, *The AAPL Guide for Landmen,* rev. ed. by Milton L. Brand (Tulsa: Kraftbilt, 1980), p. 94.

[13]Sterling v. Park, 129 Ga. 309.

[14]*23 American Jurisprudence,* par. 32.

[15]Roberts v. Abbott, 192 P. 345; Cross v. Weare Com. Co., 38 N.W. 1038.

[16]McKee v. West, 141 Ala. 531, 37 So. 740; Cannon v. Williams, 2 S.E. 2d 838.

[17]Berry v. Marion County Lumber Company, 93 S.E. 328.

[18]Harrell v. Hill, 19 Ark. 102.

[19]Willliam W. Morrow, *Spanish and Mexican Land Grants* (San Francisco: Bancroft-Whitney Company, 1923), p. 18.

[20]Gratz v. Land and River Improv. Co., 82 F. 381, cert den 173 U.S. 705.

[21]Trenton Potteries Co. v. Blackwell, 43 A. 2d 831; 137 N.J. Eq. 113.

[22]Howard R. Williams and Charles J. Meyers, *Oil and Gas Law,* abr. ed. (New York: Matthew Bender, 1975), par. 604, p. 301.

[23]Chappell v. New York, N.H. and H.R. Co., 62 Conn. 195.

[24]Williams and Meyers, par. 310.4, p. 114.2.

[25]Edwin T. Stitt, "Eastern Overthrust Belt," *Landman* (June 1982), p. 31.

[26]Howard R. Williams and Charles J. Meyers, par. 340.1, p. 193.

[27]Ibid.

[28]Smith v. Pendell, 19 Conn. 107; 2 Blackstone, Commentaries, par. 447.

[29]Knoell v. Frisco Lease, Ind., 72 F. 2d 286; Lockwood v. Ohio River R. Co., 103 F. 243, cert den 180 U.S. 637.

[30]*23 American Jurisprudence* par. 176.

[31]Hickey v. Dirks, 156 Kan. 326.

[32]Owens v. Day, 207 Okla. 341; 249 P.2d 710.

[33]Wilson v. Holm, 164 Kan. 229; Midwest Oil Corp. v. Winsauer, 159 Tex. 560.

[34]Voyta v. Clonts, 328 P.2d 655.

[35]Williams and Meyers, par. 307.1, p. 102.

[36]Carroll v. Bowen, 68 P.2d 773.

[37]122 ALR 960; *38 American Jurisprudence,* Gas & Oil, par. 28.

[38]Loeffler v. King, 149 Tex. 626; 236 S.W.2d 772.

[39]Hansen v. Ware, 274 S.W. 2d 359; Little v. Mountain View Dairies, 217 P.2d 416.

[40]*38 American Jurisprudence,* par. 31.

[41]Jilek v. Chicago, Wilmington & Franklin Coal Co., 47 N.E.2d 96.

[42]Ewert v. Robinson, 289 F. 740; 35 ALR 219.

[43]*38 American Jurisprudence,* par. 35.

[44]Davidson v. Midstates Oil Corp., 211 La. 882, 31 So.2d 7.

[45]Woodworth v. Franklin, 85 Okla. 27.

[46]*38 American Jurisprudence,* par. 19.

[47]Guffey v. Smith, 237 U.S. 101.

[48]Collins v. Abel, 151 Ala. 207, 44 So. 109.

[49]*6 American Jurisprudence,* par. 1.

CHAPTER FOUR

Who Owns This Land?

A geological department may develop a prospect for possible exploration. A company may take a "deal" from another company or individual on a prospect. The leases at this stage may or may not already have been acquired.

A plat is prepared with the prospective area outlined. This information is passed to the land department, which makes a "lease check" to determine if the area is open (unleased), already under lease to other parties, or possibly producing. If the acreage is under lease, an agreement might be made with current lessees to "farm in" the leased tracts. Such contracts are discussed in a subsequent chapter.

An area of interest is called a "prospect," and the name assigned to it may be geographical, geological, or whimsical. For example, the Northwest Potter Prospect might be situated five miles northwest of a field or town called Potter.

Fee Lands

How does the field landman begin? He will, of course, be told where the prospect is located, although the county and state may not be printed on the plat that he is given. It is advisable to print the name of the prospect, county, and state on the lower corner of the plat or on the reverse side, if only for future reference. Nothing is more frustrating than to find a plat in a file months later and wonder what it refers to. Plats also need identification as to scale. The scale is used to determine distances and size of tracts, particularly in areas measured by metes and bounds.

As soon as possible, the landman determines by a preliminary check of the records whether the lands are leased or open. He discovers who the likely parties are to execute oil and gas leases, who owns the surface, and whether any easements cross the lands.

Unless the landman is an attorney qualified to render an opinion on the title and has been requested to do so, he is not charged with actual examination of the title. He searches for "record ownership"—who, according to the chain of recorded instruments, are the apparent owners of the interests. At this stage, the landman is not concerned with "marketable" title, which refers to a seller's title beyond and free of reasonable doubt.

The statute of frauds adopted by most states provides that real estate transactions must be written instruments, executed by the granting party, and in proper form to be recorded. A handshake is not enforceable in court, and few states recognize oral agreements. The recordable instrument will contain elements of a contract. It should set out with clarity the intentions of the parties. Unfortunately, many instruments of record are not precise or easy to interpret.

The local abstract or title company (sometimes called a "plant") is usually the best and quickest source of information. The landman would first ask his employer if his company were willing to pay the fees charged for use of the plant. Some abstract companies charge a flat rate; some charge by the hour (an average fee is $15 to $25 per hour, possibly more); a few do not charge at all. Still other abstract companies do not allow outsiders to use their facilities, but they will prepare a "run sheet" or other title data for a fee.

Each abstract plant sets up its own system. The landman using the abstractor's records is advised to adhere strenuously to the rules. It has taken the abstractor many years and a substantial amount of money to compile the information. Daily use of the plant is his livelihood. The abstractor is granting a favor by allowing others to use his facility (even when he is paid for it). Avoid asking special favors of the employees, and work quietly and inconspicuously. Be friendly, and inquire for directions or clarification of the system when necessary. Also ask which records may be photocopied and if the photocopy machine may be used as needed or if the plant personnel prefer to make all photocopies.

It is wise to take notes as the system is explained. Oral instructions are easily forgotten. Several sources of indexes may be used in sequence. If the abstractor uses tract books, each tract of land will probably have all of the instruments of record listed in chronological sequence by *filing* date. Two or more books may apply to one tract. The acreage being checked may have been partitioned out of a larger parcel of land.

Many abstract plants maintain card systems, with cross references for names of individuals and all transactions in which each individual has been involved in the county. Another book or file may list all probates, court suits, divorces, marriages, bankruptcies, etc. New methods incorporate microfisch and computer printouts.

The same information may be obtained from the county records if the abstractor will not allow use of his plant or if the seeking company or individual employer does not wish to pay the fee. County records are maintained in the courthouse or records building in the county or parish seat. Tax records will be in the same town, in or near the courthouse, as will probate, death, marriage, and divorce records. The district court covers more than one county and may be located in another nearby town.

The prospect plat may have names of individuals printed on the most southerly side of the tracts. If so, it has probably been shot from an "ownership" map. Ownership maps are prepared by various companies (such as Tobin, Heydrick, and Pomco). The maps are protected by copyright laws, but companies that have purchased certain base maps may have permission to photocopy it by areas. Many states, including West Virginia, Maryland, and Pennsylvania, have no ownership maps available.[1] In this case, the landman is forced to compile his own map with the aid of tax and courthouse records.

Names on a plat are good leads to begin a search of the records. However, the names may not be current. The search for present ownership begins at the tax assessor-collector's office. Tax plats, if available, are generally indexed by tract numbers. Often there are no tax plats. Tract books in the tax office are maintained by years and indicate the taxes rendered and paid on each parcel. Major headings note township and range or survey and abstract number. Names of persons assessed each year are the first clues in locating the surface owners. Many states do not assess minerals separately until an area produces.

Because of the oil play that began in Pennsylvania in the mid-1800s, severances in that state may date back more than 120 years. In the Eastern Overthrust area, as early production ceased, severed mineral interests broke into smaller and smaller fractions by the process of intestate inheritance.[2] The titles are tricky. According to Edwin Stitt, "A West Virginia statute provides that a tax sale of realty will pass all the right, title, and interest of the owner in default, plus all other interests therein that have not been assessed separately."[3] This means that mineral rights that may have been severed from the surface but not assessed for tax purposes could pass by tax sale to a new owner. Pennsylvania distinguishes between "seated" (occupied) and "unseated" (unoccupied) lands. A tax sale of seated land would act only as a sale of the interest that was owned by the person in default. It would not include the severed minerals.[4] The landman should familiarize himself with the characteristics of real estate and probate law in the state in which he is working.

Occasionally, tax records indicate the severance of minerals. A notation of the names of owners and years in which changes appear will be useful in finding deeds of transference.

For the lease check (sometimes called a "lease takeoff"), the landman is concerned with oil and gas leases currently in force that cover the acreage under search. Oil and gas leases may be included in deed records or may be compiled with oil and gas records or contract records. A person on the staff of the county clerk's office will explain and instruct regarding the system. The names of county clerks' offices vary from state to state—clerk of court, court registry, etc. For simplicity, however, the term "county clerk" is used in this text. Probate or surrogate court records and deed records may be in the same office or in different offices.

Indexes are chronological and alphabetical by name of grantor and grantee. The current direct (grantor) index is a starting point to determine if the names of the tract owners appear to have granted recent oil and gas leases. Indirect indexes are alphabetical by name of grantee. Many states also have tract indexes. In such states, the parcel may be examined by location, as from an abstractor's records. Instruments are listed in the index according to filing date (as opposed to instrument effective date), although the date of the instrument is frequently noted. Filing date is important for purposes of constructive notice. Should a lessor execute two leases on the same tract of land, the first to be recorded is the valid lease, regardless of instrument date. The date of the instrument is critical for terminable grants in order to determine expiration dates. Record searching requires diligent thoroughness. Hugh L. Scott, attorney and landman from San Antonio, Texas, likes to say, "If you don't have time to do it right, will you have time to do it over?" The last thing a landman wants to do is buy a lease from the wrong party or pass wrong information to his client or employer regarding the leasehold status.

The name of the grantor may be followed in the direct index to find the volume (book) and page on which the instrument is recorded. Lucky the landman who discovers a brief description in the index that indicates the lands covered by each instrument. The descriptions—if any—are not all-conclusive. It is possible that there are second or third tracts in the same conveyance that are not noted in the index. It is necessary to read each instrument carefully. A grant in one paragraph may be qualified in another.

If the index does not contain a description of the lands covered by each instrument, the conveyances must all be checked by grantor-grantee names pertaining to the tract in question. It is easier to work backward. Going back 15 to 20 years for a lease check may be sufficient to determine if the tract may still be producing under an old lease. In old producing areas, it is advisable to check back even further. Releases and assignments of leases, as well as unit agreements and dissolutions of unit filings, are important.

Instruments recently received by the clerk for filing may not yet be listed in the current index. The clerk maintains a "file register" or list or box of newly received items. These may be checked daily to stay abreast of activity. Unfortunately, many counties may be months behind in posting.

Map and plat records are separately indexed. A review of all of the plats available on the parcel is enlightening. Plats are valuable for names of owners through the years. In areas not under governmental survey, names of surrounding tract owners assist the landman in locating a parcel that has been carved out of a larger tract.

Occasionally, the original work plat given to the landman will indicate the locations of well sites. Old units may remain in effect. The state office that regulates drilling activity is helpful in furnishing information regarding well status—whether a well is producing, plugged and abandoned, or shut in. The county surveyor, whose offices are generally in or near the courthouse, has a wealth of information and maps on well locations, boundaries, easements, and other data of benefit to the landman and landowner alike. Although it is generally the landman who checks records, all of this same information is available to the landowner who is interested in whether adjoining lands are leased or whether a permit has been granted to drill a well in the vicinity.

Discretion on behalf of one's employer is certainly advisable to the landman working in public records. An area of interest derived from a geological prospect may be worth a substantial sum of money. Ethics demand that the landman hold his knowledge in trust. He should avoid sharing the data with the person standing nearby by carelessly leaving a plat or run sheet where it may be seen or copied.

Good-quality colored pencils, a straightedge, and a light hand are necessary to prepare a neat plat showing the "lease takeoff." A separate color may be used for each lessee. For instance, green might denote Texaco leases, orange the Johnson Oil Company, blue for Exxon. Yellow is customarily reserved for one's own company. Red is the general outline color for the area of interest, but no rule applies. Units are outlined in contrasting colors. The names of the lessees and the expiration date of the lease may be printed across the inside most northerly area of each tract. The landowner or mineral owner's name may be inserted on the lower inside area of the parcel. If several persons own undivided interests, their names may be printed with a fine-point black pen where they will be most legible and an arrow used to designate the tract the information goes to. Partially leased tracts may be crosshatched. Unleased tracts might remain uncolored, with the word "open" printed on each. A good landman learns to be a fair draftsman.

The report that goes with the plat may be typed or written on a pad or a form designed for the purpose. Headings for the needed information include Lessor, Lessee, Instrument Type, Instrument Date, Volume/Page, Brief Description, Remarks. A separate report is made for each tract. The data are attached to the plat and submitted to the employer or company. A decision may be made either to proceed with a leasing program or to try to farm in the acreage from present lessees.

A more extensive method of record searching is used to check records before buying a lease. An accurate appraisal must be made of the exact name, address, and fractional interest of each mineral owner. This is called a "mineral takeoff." The mineral estate may be severed from the surface by conveyance, reservation, litigation, or heirship. Lis pendens (suits pending) and county and district court records reveal litigation and partitions. Divorce proceedings are usually included in district court records.

Death records are not always available for search by the public. A request of the clerk to search and provide a copy of the death certificate might be honored for a small fee. In large cities, death records may be found in city or in county health records, depending on the site of death (not always the place of domicile). Probate records are separate from deed records but are a part of the county record system. If no will has been admitted for probate, a search of the deed records might reveal an affidavit of heirship. Marriage records are included in the county records.

Deed records consist of transcribed copies of original instruments. These were copied by hand or by typewriter until the photocopy machine came into existence. A transcriber was honor bound to copy the instrument verbatim, regardless of error in the original. Deviations are called to the reader's attention by the use of a line under the error, e.g., marksd.

The landman is accountable for locating the tract on the ground with his plat. He must be certain the description he is reading is the tract he is tracing. Governmental survey descriptions all begin to look alike at times. Metes and bounds property descriptions, particularly in older deeds, may refer to trees marked "X," meandering streams, monuments, or iron pins. A description may be by deed reference, i.e., a recording reference to a prior deed of record. All deed references should be checked for accuracy. Circuitous descriptions and ambiguous calls may be compared with other deeds of record or correlated with plats to confirm the location. It is recommended that the landman drive to and personally inspect the land he plans to lease. The local chamber of commerce will supply a county road map.

A list of the mineral owners is prepared from the chain of title. Addresses of the owners and the fractional interest owned by each are included. The fractional interests must add up to 100% for each tract. Now and then, overconveyances will indicate more than 100%. A diagram that indicates the chain of conveyance is very helpful during all stages of the title work.

Once the mineral ownership is established, prospective lessors may be contacted for leasing.

Federal and State Lands

With the exception of the original 13 states, Texas, and Hawaii, each state acquired lands from the federal government as it passed into statehood. In the early days of the Union, states that entered the Union would convey their western lands to the United States. Georgia and Texas were exceptions. Federal grants were frequently made to the states under provisions of "selection," whereby a state could select its choice of certain federally owned acreage within its boundaries. The General Land Office of the United States issued a federal certificate indicating that the state was granted title. Selection lands included school lands, navigable rivers, and tidelands.[5]

Federal land records are kept in the office of the Bureau of Land Management in the area of jurisdiction or in Washington, D.C. Serial registers and title plats are available to show the current ownership and records of past transactions. Case files hold original documents. The landman is not generally expected to go to Washington, but he might be asked to look into regional BLM records.[6] A number of Washington, D.C., attorneys specialize in federal land titles.

Information regarding the title to state lands is retained by the office of the state land commissioner or other state agency handling land matters. By prior agreement with England, the original 13 states had acquired public lands, which they retained. Texas and Hawaii had been independent nations prior to statehood, and each retained its own public lands.

A patent from a state is considered original or sovereign title. It is not necessary to search behind a patent. Any patent from the federal government or from a state should be examined to determine if the governmental authority reserved the mineral rights or other rights that would affect either the lease or development.

Indian Lands

An excellent discussion of the history of titles to Indian lands may be found in "Indian Lands" by Gary Boyle in the October 1981 issue of *Landman* magazine. Boyle sets out five sources to be searched in determining Indian titles:

1. Federal laws (treaties, statutes, executive orders)
2. Appropriate regional BLM office
3. Bureau of Indian Affairs title plant covering the area
4. Bureau of Indian Affairs, area and agency offices
5. County clerk's records[7]

Indian titles, whether to tribal lands or alloted lands, are subject to original grants from the United States government. Minerals are subject to the reservations, restrictions, and laws against alienation and to term limitations. Searching title to Indian lands is a specialty that deserves careful study and consideration prior to making a bid for a lease. Rules and regulations may be obtained from the Bureau of Indian Affairs. The commissioner of Indian affairs in Washington, D.C., and area directors hold the title information.

References

[1]Edwin T. Stitt, "Eastern Overthrust Belt," *Landman* (June 1982), p. 20.

[2]Ibid., p. 28.

[3]Ibid., p. 29.

[4]Ibid., p. 30.

[5]Lewis G. Mosburg, Jr., *Landman's Handbook on Petroleum Land Titles* (Fort Worth: Institutes for Energy Development, 1976), p. 5.

[6]Leslie Moses, *The AAPL Guide for Landmen,* rev. ed. by Milton L. Brand (Tulsa: Kraftbilt, 1980), p. 39.

[7]Gary Boyle, "Indian Lands," *Landman* (October 1981), p. 21.

CHAPTER FIVE

The Oil and Gas Lease and Leasing Procedures

Fee Leases

Oil and gas lease forms developed soon after the discovery of oil at Titusville, Pennsylvania, in 1859. Early leases required either development or forfeiture, and primary terms were for specified periods of time. There were no delay rentals, and cash bonuses were not paid. Development served as the remuneration for the granting of the lease. Leases that were silent regarding specific development obligations brought court decisions holding for the "implied covenant" to develop.[1] Usually the leases covered small surface acreage, and one lease would be developed before another lease was acquired.[2]

Situations evolved wherein a lease that had no provision for extension would terminate, even after production had been found. Thus this phrase came into usage (directly following the primary term): "and as long thereafter as oil and gas may be produced from said land or land with which said land is pooled." This is referred to as the "thereafter" clause, which is more fully discussed on pages 56–57. The primary purpose of the oil and gas lease is to allow the lessee to develop the leased premises without a specific obligation to do so; if production is discovered, the lessee needs the right to maintain the lease for as long as is economically feasible.[3]

Many lease forms are called "Producers 88." The designation is said to have originated with the Burkhart Printing Company in 1916. Burkhart printed leases for the Producers Oil Company, using the word "Producers" for his client and the number 88 as his own numerical system.[4] Several stories exist concerning the development of the Producers 88 lease. Such a variety of revised Producers 88 forms exist today that the label cannot be construed to indicate any consistency of provisions. Hundreds of differing printed lease forms are in circulation. It is not unusual for an attorney to prepare a special typed lease for his client. The most widely used form of lease today evolved from an Oklahoma landowner's suit to cancel his lease (Brown v. Wilson, 58 Okla. 392, 160 P. 94), after which several major oil company attorneys agreed on standard provisions in the drafting of a new lease form.[5]

While there are no standard leases, there are standard provisions, with variable language to convey the intent of the parties. Special clauses are often attached as riders or in an exhibit to suit individual circumstances. The prudent landman is alert for provisions out of the norm. He is responsible first and foremost for reading word for word the lease that he is negotiating and buying for his company or client. He should understand the terms and be able to explain their significance to the prospective lessor.

The landman is the company's (and often the industry's) first contact with the landowner. As the adage says, there is only one chance to make a good first impression. The supersalesman is out of place trying to buy an oil and gas lease. A straightforward, forthright approach usually ensures confidence and opens the door for negotiation. The good landman strives to be a fair trader as well as a good negotiator for his company. A "deal" is satisfactory only when it benefits both parties. Outsmarting a landowner in a way that will later lead to a bad relationship could easily preclude future critical dealings with that party or with his neighbors.

Landowners are not necessarily cordial. The landman's reception can be rude or hostile, and he must take these attitudes carefully in stride. Certainly he must resist a retort that might hinder further conversation and discussion. But no matter how diplomatic or how knowledgeable a landman might be, a lease cannot be bought if the lessor (1) doesn't want to lease under any circumstances, (2) refuses to discuss the matter, or (3) sets terms that are uneconomical or unreasonable for the proposed lessee.

The key to buying a lease is to find out under what circumstances the mineral owner is willing to lease, reconcile these with the needs and terms offered by the lessee, and bring the two to a reasonable compromise in order to satisfy the needs of both parties. The landowner who initially says "No" may be unaware that there are options to work out the terms. He may fear that a well will be drilled in his vegetable garden or too near his home or that his cattle or fences will be destroyed. The experienced landman, attentive to the landowner's needs, offers solutions by way of special clauses or patient explanation of lease terms that the prospective lessor might not understand. The lessor may desire a deferred bonus payment for tax purposes. In this case, a letter agreement should set out such terms of delayed payment and should state that the lease is in full force and effect from its effective date.

Primary term, royalty, and bonus per acre and delay rental are the most commonly negotiated points of the oil and gas lease. The initial bonus offer is set by the company buying the lease. Economics is the basic of the business. The lessee will examine the overall cost of the prospect—leasehold costs, maintenance, possible renewals—along with drilling and development costs. Most companies attempt to buy all leases in a block at one bonus and royalty for the same primary term. The royalty offered to the landowner is based on the going rate in the area. The more expensive the well, the lower the operator hopes the royalty and other burdens against net revenue interest will be so he can recover well costs within a reasonable period of time (payout). Payout is calculated on anticipated production, well cost, market price from purchasers of production, costs of getting the product to market. Is the location inaccessible, requiring expensive roads? Is there a pipeline in the area? Is there a purchaser for the gas? Will costly and time-consuming rights-of-way need to be acquired and the cost of a pipeline assumed in order to get the product to market? Will the gas need to be treated? Not all answers are known factors at the time the leases are purchased. From the landowner's standpoint, what is the going rate in the area? Are there competitors who might make a more attractive offer? Does the prospective lessee plan to drill, or will he hold the acreage for speculation?

Face-to-face negotiations afford the opportunity to look one another in the eye and assess one another's stance. A written proposal, with a copy of the lease, might routinely be offered to a company, bank, church, governmental authority, or other entity; but personal contact by the professional landman is more effective than the telephone or mail.

When all offers in an area are identical and tax records or county clerk's records are up-to-date (with town lot leasing, for example), it often saves time to take along the lease and a draft for the bonus on the first visit. However, each situation warrants its own approach. The lease may be prepared after negotiations are concluded. By this time the landman knows the proper styling of the lessor name or entity, correct address, acceptable terms, name and address of depository bank, and special provisions that might be needed.

Lessor-Lessee/Effective Date

All necessary information should be typed in the blank spaces on a printed lease form. A landman who is not an attorney is not qualified or legally entitled to prepare a lease or similar contract from scratch. Changes in the printed form should be initialed by the lessor at each change. Although certain state statutes require a lease to be signed by both lessor and lessee, it is most frequently executed only by the lessor ("lessor" being used for both singular and plural).

The effective date of the lease (unless otherwise provided) is the date the agreement is made and is indicated in the space at the top of the form. The actual date of execution is not the determining date. References in the body of the lease to "this date" are to the date on which the agreement is made, i.e., the date typed into the contract. In a situation where a lease is filed of record and the date has been inadvertently omitted, it would be appropriate to have an amendment to oil and gas lease executed so that the effective date could be properly established and the intent of the parties would be a matter of record. Otherwise, the earliest acknowledgment date would likely prevail as the date of contract.

Occasionally, as in a top lease, the effective date of the lease is not the same date that the agreement is made. In this case, the effective date should be distinguished from the date of the instrument. This information is most frequently inserted in the habendum clause after the statement of primary term, i.e., "for a term of five (5) years from *March 3, 1982.*" The words "from the date hereof" would be stricken in favor of the insertion of the effective date. The effective date may also be inserted at the top of the page by the date of agreement, with the words, "This agreement made *to be effective* this *3d* day of March." At the end of the lease instrument, above the signature lines, following "this instrument is executed on the date first above written," clarification is assured by the addition of the words, "but effective as of March *3d*, 1982."

The lessor name properly conforms to the name in which he holds record title. If he prefers his name written differently, or if the name has changed by marriage, the phrase "also known as (aka)" (whatever the record owner name is) can be inserted after the name by which he wishes to sign. The point is to conform with the record owner name, one way or another, so that when the instrument is filed there will be no doubt about whether the person called John J. Smith is one and the same as John Joseph Smith or J. J. Smith.

An unmarried lessor is designated "a single man" or "feme sole" or possibly "widow" or "widower," following the typed name. A married person dealing with his separate property would generally be styled, "John J. Smith, his separate property, being no part of his homestead." If the spouse joins in signing, for community property, homestead, or other reason, the designation of spouse clarifies the relationship: "John J. Smith and wife, Martha Ann Smith." Without the spouse designation, future readers of the recorded instrument will have doubt whether Martha Ann Smith was his wife, daughter, mother, sister, cousin, niece, or other relative.

Married women properly sign with given names. Combinations of names in the heading are techniques that save curative time and often avoid the necessity for affidavits of identity. For instance, Jane J. Jones might be shown as "Jane J. Jones, aka Jane Johnson Jones"; or if she were divorced, as "Jane J. Jones, formerly known as Jane Rogers Johnson, aka Mrs. John L. Jones."

Individuals signing in a fiduciary capacity for another are indicated as "John J. Jones, Trustee (Administrator or Executor) for Ann Crawford Smith" or "John J. Jones, Guardian for the minor, Robert Price." The designations of guardian or trustee without the name of the beneficiary create doubts. There are blind trusts, however, in which the trustee will not divulge the beneficiary and will not furnish a copy of the trust instrument. The vast majority of trust agreements are unrecorded documents. Trusts created under the terms of a will are, of course, public record. If these terms are not available in the county where the leased premises are located, it is advisable for the landman to furnish copies of the items with the lease he has taken in order to substantiate the ownership. This also applies to powers of attorney. The latter executes the lease as agent and attorney-in-fact for the lessor.

The complete address, including Zip Code, follows the lessor's name. Likewise, full address of lessee appends its name.

Consideration/Granting Clause

Several types of consideration exist—good, valuable, and illegal (inadequate or prohibited by law). A fundamental principle of contract law is a mutuality of obligation.[6] The consideration given by the lessee to the lessor for an oil and gas lease must be in accord with the prevailing state statute.[7] In Louisiana, for example, if no money consideration is included, there must be an unconditional agreement to drill and develop the lands.[8] In other states, a promise to develop or to pay rentals might meet requirements. A "valuable" consideration should be deemed in fair relationship to the worth of the exchange. Inadequate considerations for the purpose of setting properties out of the reach of creditors can be set aside by the courts as illegal.

The oil and gas lease commonly uses the term, "ten dollars and O.V.C." (other valuable considerations) to avoid placing of record the exact amount of the bonus money involved. This protects the lessor by allowing him the privacy of his transaction and protects the lessee in the event that negotiations of varying amounts are transacted within one area. Good or valuable considerations may be money, exchange, or obligation, depending upon state law.

The granting clause "grants, leases, and lets exclusively unto Lessee for the purpose of," succeeded by all-inclusive descriptions of the anticipated rights of the lessee that are consistent with exploration and development operations. These rights are broad and include rights of ingress and egress for geophysical surveys, building roads, laying pipelines, and transportation and structural needs. It does not follow that all of these rights will be utilized in the course of operations. Provisions may be limited by the inclusion of words, "*necessary* to Lessee in operations to produce, save, take care of, treat, transport" or otherwise carry out development procedures. If the landowner is concerned about the use of his surface, particular drillsites or restricted areas may be agreeable to both parties to the lease. The lessee has the right to drill within the vertical boundaries of the surface of the leased premises. He may not slant drill to bottom the well under a tract of land owned by another mineral owner, even if he owns the lease, unless he is granted a permit for a directional well by the controlling governmental body.[9]

Description

A proper legal description should allow exact location of the land on the ground. The governmental survey system used by many states precludes ambiguities when the description is properly written. Care must certainly be taken to avoid typographical errors with sections and quarter sections, townships and ranges, blocks, lots, and subdivisions.

Metes and bounds descriptions are used in many southern states, where old land descriptions have been handed down from French, Spanish, and Mexican grants. Patents were issued on old field notes that later proved inaccurate. Consequently, surveyed tracts did not always "come together" at the boundaries, and vacancies were created. Some descriptions go from monuments or points such as trees, iron pins, or the meandering of streams that may no longer exist. Still another description used in the early part of this century located a tract by designation of adjoining landowners.

With an engineering scale, the landman or landowner is advised to track his metes and bounds description on a plat to assure himself of the location of the lands. Deed reference descriptions are commonly used, but care must be taken that the deed referred to has a proper metes and bounds description or field note survey. The deed reference description should not refer back to another deed reference.

A brief description of the leased lands may leave a space that can be used for the insertion of conditional clauses. A legal technique to prevent insertions or alteration of the description is to draw a line in the design of a "Z" from the end of the description to the next printed line.

A description too lengthy to fit into the space provided may be typed on a separate sheet of paper (preferably white bond), with the designation "Exhibit A." The words, "See Exhibit A attached" would be typed on the face of the lease instrument where the description goes. If exhibit pages are attached to a printed lease form, each page of the lease may be numbered as "Page 1 of 5" to "Page 5 of 5." If the pages become detached, the person handling the instrument is alerted. For an Exhibit A, the following wording is recommended:

> Attached to and made a part of that certain Oil, Gas and Mineral Lease dated __________, 19____, from John J. Smith et ux, Jane Smith, Lessor, to Hoopla Oil Company, Lessee, covering 111.11 acres of land, more or less, situated in the S½N½SW¼ and S½SW¼, Sec. 9, T8S-R9E, Monroe County, Mississippi.

Each page would have a line at the end of the description to be "Signed for Identification."

Fly sheets are commonly attached to leases and taped on the face of the lease where the description would have been typed. This kind of addition to the form also needs a "Signed for Identification" line for the lessor's signature. This form is less desirable than an exhibit. Certain counties charge an extra fee to record an instrument with a fly sheet attached; some prohibit this form for recording. There is always the danger the sheet will become detached and lost.

Mother Hubbard

A "Mother Hubbard" or "cover-all" clause is included after the description in many states where inaccuracies have abounded with metes and bounds descriptions. Courts have construed this clause to "bring within the conveyance small strips of land bordering the described tract . . . which may not be included because of faulty description or may have been acquired by adverse possession."[10] Its purpose does not expand the intention of the parties. The cover-all phrase originated to provide for small strips of land affected by offset well provisions.[11] An improper description might deprive the landowner of benefits or royalties from acreage that failed to be specifically described. The Mother Hubbard acts to include adjacent, adjoining, or contiguous lands that might be omitted due to improper description, inaccurate survey or vacancy, or adverse possession of which the landowner may or may not be aware.

On the other hand, an astute landowner might strike this provision or be certain that he is in accord with court definitions:

> adjacent: near to, in the neighborhood (can be across the street)
> adjoining: touching at some point
> contiguous: all or most of one side in common[12]

It is important that rental or shutin payments not be altered by the inclusion of small strips of land covered under the Mother Hubbard. For this reason, a set amount of acreage on which rental payments will be based is inserted in this clause. In the event of unit production, royalties are based on actual surveyed acreage.

Habendum Clause, Primary Term

The primary term of a lease is designed to provide sufficient time for the lessee to assemble his lease block, conduct and evaluate geophysical tests, determine the appropriate drillsite, obtain a rig, prepare the location, and arrange for necessary drilling services. In Louisiana and Tennessee, a primary term may not exceed 10 years.[13] Otherwise, it is a term mutually agreed upon by both parties. Printed lease forms generally specify 10-year primary terms. At this writing, the majority of fee leases are taken for 3- to 5-year terms. One-year or 6-month leases are not uncommon. Ninety-day to 180-day leases serve almost as a farmout. The short-term lease assumes prompt development.

The "thereafter" clause in the habendum assures the operator that the lease will remain in effect in the event of *production's* (or at times, *operations'*) occurring across the expiration date. Once the primary term is passed, the lease held by production (HBP) from its own premises, by unit production, or by other means is said to be into its secondary term. Temporary cessation of production due to justifiable conditions after the primary term will not terminate leases that contain a "saving clause," which is discussed later in this chapter.

Lessor Royalty

Royalty is a provision for continuing remuneration to the lessor.[14] Royalty for oil is distinct from royalty for gas, although the two typically carry the same fractional interest, as do associated hydrocarbon products. Sulfur and other minerals might characteristically be one-tenth or 50 cents to $2 per long ton. The latter rarely become items for negotiation. A landowner might consider this point in his dealings, for he is probably better served by a fractional interest than a dollar amount. Oil and gas royalties are the "meat" of the contract.

Unless a lessor has a need for immediate cash bonus monies, he would do well to consider the long-range benefits of a good royalty provision. Royalty income is not assured. Perhaps 14 out of 15 wells drilled across the country will be dry holes. However, in the event of production, royalty can bring substantial income to the owner. Bonus is the bird in the hand; royalty is a gamble for higher stakes. There is a substantial difference between one-eighth royalty and one-sixth royalty, as illustrated in the following example:

A completed well produces 300 barrels of oil per day (BOPD).
At $30/bbl × 30 days per month, gross income = $270,000/mo
  Lessor royalty @ 1/8 = $270,000 × 0.125 = $33,750/mo
  Lessor royalty @ 1/6 = $270,000 × 0.166 = $44,820/mo

This is a simplified example and does not take into account the taxes attributable to royalties. However, it illustrates the potential value and importance of the royalty provision, for royalty continues during the producing life of the property. Long-term income may provide a tax advantage over the one-time bonus.

Lessor royalty is subject to outstanding nonparticipating royalties of record. In the event of pooling, royalty is proportionately reduced according to surface acres in the unit; this is more fully explained under the pooling clauses that follow.

Critical to the royalty provision in the lease is the wording on which the market value of the product is based. "Fair" market value has been the subject of extensive litigation. The most famous case of record establishing fair market value is the Texas Oil & Gas v. Vela, 429 S.W.2d 866, better known as the Vela case. The lessor contended that fair market value should be computed in accordance with the open market price then available for the product in lieu of the actual price received by the lessee under a long-term gas contract. A ruling in favor of Vela was upheld by the Texas Supreme Court in 1968. However, in 1981, the Texas Supreme Court held that lessor's fair market value gas that was marketed through interstate pipelines could be calculated with reference to federally dictated prices rather than with reference to the open market prices in the area (First National Bank v. Exxon, 622 S.W.2d 80). The latter decision was justified by a differentiation between the "quality" of interstate gas as opposed to intrastate gas in the Vela case.

Fair market value in the area can be construed to include relatively large areas of several counties or districts. Fair market value in the field is easier to determine by being more restrictive. Market price at the wellhead might assume the actual price paid to the operator by the purchaser as it comes from the well. Market price at the pipeline can entail treatment and transportation charges. Conflicting court decisions in various states have created technical differences between market value and market price, with an obligation on the part of the lessee to obtain the best contract available.

Shutin Gas Royalties

Payments of shutin gas royalty are generally provided in the oil and gas lease as remuneration to the lessor to keep the lease in full force and effect for periods of time after production is found and before a market is obtained for the gas or gaseous products. Other situations warrant the shutting in of a well, including testing or workover procedures, but these are likely to be covered under the saving clause for resumption of operations. Careful review of shutin provisions is necessary, for each lease may be phrased differently.

The shutin clause was first used in the 1930s and became more common in the 1940s and 1950s.[15] Even though the payments have the characteristics of rental payments, they are called royalties and take the place of royalties paid on production.

The timely payment of shutin royalties is critical. Most leases provide 60 to 90 days after shutin before the payment is due, with ensuing payments to be made on certain anniversary dates of the first payment. Amounts of the payments vary from small sums (in older leases) to sums equal to annual rentals or even larger sums for a restricted period of time. Certain state statutes consider shutin payment as an advance on royalty. The majority of leases provide that the shutin payment is in lieu of drilling operations and that when such payments are made, "it shall be considered that gas is being produced in paying quantities." For a producing well status, Texaco, Inc. v. Fox, 618 P.2d 844 (Kan. 1980), defined "paying quantities" and "commercial quantities" as having the same meaning and ruled that depreciation shall not be used as an expense in determining paying quantities.

Failure to make a timely shutin payment is tantamount to failure to make proper and timely delay rental payment and creates a forfeiture of the lease. There is a common saying that you cannot breathe new life into a dead lease. The only remedy for nonpayment would seem to be to approach the lessor with cash or check in hand, and pray that he is receptive to negotiation. The first shutin payment and each anniversary payment are important factors in this clause. Monthly shutin gas payment requirements should be avoided if at all possible. The paperwork and bookkeeping can be a headache, and the margin for error is increased 12:1. Many leases are limited in the time period by which shutin royalties will hold the lease. This stipulation may be added by special clause if not included in the body of the printed instrument. Two to three years is a reasonable and normal length of time to hold a lease by shutin payments.

Other Minerals

Sophisticated mineral owners, particularly those who own the surface of the leased acreage, frequently delete any reference in the oil and gas lease to "other minerals." Oil and gas are construed to mean hydrocarbons and related substances, excepting coal. The coal lease has come into existence to cover the rights of the surface owner to coal that, if strip-mined, would destroy the surface. Often a coal lease is taken from both the mineral and the surface owners. "All other minerals" are generally considered to include hard rock minerals and ores, including fissionable materials. Gravel is considered as belonging to the surface estate.

Pooling Provisions

Pooling clauses constitute fairly recent efforts of the oil industry toward conservation and efficiency of operation. They also allow conformance with spacing rules and/or zoning regulations under the applicable state jurisdiction. Rarely do spacing rules cover authority to include federal or Indian lands, which are subject to their respective regulations.

"Pooling" may be the joinder of two or more small tracts to comply with well spacing obligations. Pooling is an attempt to avoid economic waste by allowing greater recovery at less cost, by avoiding the drilling of unnecessary wells, and by protecting correlative rights of the small tract owners.

The basis for authority that the lessee has to pool his tracts is either explicit language agreed to by the lessor in the oil and gas lease or compulsory pooling under statute. Pooling for oil is most frequently done on a 40- or 80-acre basis. Gas units range from 160 to 320 to 640 acres. A 10% "tolerance" allows a necessary variance factor. Some wordings leave the size of the unit to be determined by the governmental authority that has jurisdiction. The words "prescribe or permit" within the pooling clause refer to pooled units that may be created by order, in accordance with conservation laws and allowables, or are permissible under regulations.

The outline of the unit is usually based on surface acres, although the pooled acreage may be comprised of one or more subsurface strata. Units may overlap if they cover differing horizons. The Louisiana Conservation Commission determines its unit configurations upon subsurface geological data. No minimum portion of the leased acreage must be included in a pooled unit unless so specified within the lease.

In most states acreage is considered pooled at the time the lessee prepares and files an instrument of unit designation for record in the county where the land is situated. Louisiana requires a hearing before the Conservation Commission. The unit designation usually includes an exhibit of all leases to be included in the unit and may or may not have a plat attached. In most states the unit designation may be filed before or after the well is drilled if the lease so allows. It becomes effective as of the date of filing or other date specified therein. The power to pool may be exercised by the lessee time and time again, although modification of existing units is not generally allowed. Properly, if the unit fails to produce or production ceases, a dissolution of unit should be filed of record by the lessee or appropriate owner of the leasehold interests. Old unit designations that have never been dissolved are of record in countless courthouses.

The entire acreage in a pooled unit, except for the payment of royalties, is considered as one lease. Operations or production from one tract in the unit is considered to be operations or production from all tracts. Unless there is a provision to the contrary, operations or production on the pooled acreage will hold all of the leased acreage covered by each of the pooled leases.

Pooling does not alter the ownership of the lessor's interest insofar as he may be entitled to delay rentals or shutin royalties under the terms of his lease. Royalties from production are allocated in proportion to the acreage contributed to the unit (tract factor) and in accordance with the royalty provision in each lease.

Communitization, voluntary and compulsory pooling, and unitization will be discussed in a subsequent chapter.

Delay Rentals

The paid up lease form does not require delay rental payments. This fact creates the "paid up" status. Delay rentals provided in other lease forms are the means by which the lease is preserved in effect from year to year in the absence of drilling or operations.

Two general types of delay rental clauses are in use: the "or" clause and the "unless" clause. The "or" clause requires specific action on the part of the lessee. He must commence drilling or pay rentals (a "drill or pay" clause); or the wording may include "surrender" of the lease. In any event, the lessee must do one or the other. Failure to do either may create an obligation to pay the rentals as damages.

"Unless" leases are commonly used. They provide that "unless" drilling (or operations) is commenced before the anniversary date or rentals are paid, the lease shall terminate. This requires no action on the part of the lessee should he elect to allow the lease to expire by nonpayment of rentals.

Courts have consistently held for strict construction of the delay rental provision. Payments are due on or before the specified anniversary date, although a few states allow a grace period. Timely payment is critical to the life of the lease. For this reason, most rental payments are made from 15 to 60 days in advance.

Depository banks are designated in the majority of leases in order that the lessee may make the rental payment by check or draft directly to the bank for the credit of the lessor. Rental checks often have an attached receipt that is signed and returned to the lessee for his records, evidencing date of payment. It is not necessary that the bank credit the lessor's account before the due date, but it is necessary that the bank receive the payment on time.

Leases have been held terminated in the event of late payment, insufficient payment, payment to improper persons, payment to the wrong bank, or late payment that was improperly addressed.[16]

The proper sum of rental to be inserted in the face of the lease is the gross acres covered by the described tract of land multiplied by the dollar per acre rental amount. The "lesser interest" clause provides for proportionate reduction of the rental and royalties in the event that the lessor owns less than the full amount of mineral interest. Care should be taken that there is a lesser interest clause in the lease. If a specified nonreducible dollar amount of rental is agreed upon between the parties, the phrase in the lesser interest clause pertaining to delay rentals is stricken.

Leases that provide for payments to be made to a depository bank assume use of the mails for that purpose if not specifically designated "by mailing." Payments to banks may usually be by check or draft. Unless otherwise stated, a payment direct to the lessor is properly made in money order, certified check, or the like if no authority to make this payment by check or draft is set out. Direct rental payments to the lessor should be made by certified mail, return receipt requested, and closely monitored to confirm prompt and proper delivery.

In the event of assignment by the lessor of his interest or a portion thereof, the lessor is obliged to notify the lessee. If the lease contains a "notification clause," the lessee is generally not in default for improper payment until after 30 days from the date of notification by the lessor of change of ownership. This also applies to changes in the lessor's depository bank. In the event of the lessor's death, rental payments are made to his estate at the designated depository bank until succession of title is furnished to the lessee in the form of probate proceedings or affidavit of heirship.

No alteration in rental sums or payments is authorized without notice in writing from the lessor to the lessee. Several lessors named in one lease may elect to execute a rental stipulation or rental division order, setting out the proportion of rental monies to go to each at his own depository bank. The burden is on the lessor to notify the lessee of any change in ownership, address, or depository bank. Until such notice or proof is furnished, the lessee may rely on the ownership indicated by his lease.

Bonus considerations are not considered to be part of the rental payment. Most leases specifically state this fact. A paid up lease might be acquired solely with the "front end" money as bonus or with delay rentals paid "up front" also. If the one-time payment includes rental monies, the check or draft should so indicate for benefit of the lessee's accounting. Bonus considerations are capitalized, while delay rentals are expenses. The paid up status is negotiable.

Release

Partial or full release of acreage may be made by the preparation of an instrument (release of oil and gas lease or partial release of oil and gas lease) that describes the acreage to be surrendered. This instrument is placed of record in the county or parish where the land is situated. In this manner, the lessee is absolved of future obligations and rental payments on the acreage dropped. Some leases provide for horizontal as well as vertical release. In the event of horizontal release of designated strata, delay rental would be due in full on the basis of the surface acreage retained.

Partial releases of acreage covered by the lease are provided under most forms, with proportionate rentals to be paid on the acreage retained. In cases where partial assignments of leasehold acreage are made by the lessee, the courts have held that the failure of the assignee to pay rentals on his proportionate part may cause a termination of the entire leased area. Lease forms used widely in Oklahoma, Kansas, Texas, and Colorado contain provisions that the termination of segregated portions by nonpayment of rentals does not affect or terminate the portions on which rentals are paid.

Saving Clause

Relief from rental payment is provided by the drilling of a dry hole that is completed within 60 days of the anniversary date of the lease. What constitutes a "dry hole" is the subject of numerous litigations. Is it "dry" if it is uneconomical to produce? What constitutes "production in paying quantities"? Production in paying quantities relies on the prudent operator standard (Clifton v. Koontz, 160 Tex. 82, 325 S.W.2d 684).[17] Production obligates the lessee or his designated purchaser of the product to pay royalty to the lessor.

The saving clause is defined in *Black's Law Dictionary:* "In a statute, an exception of a special thing out of the general things mentioned in the statute . . . intended to save rights . . . ."[18] Under the oil and gas lease, the saving clause "saves" the leasehold rights for the lessee under special circumstances. In the event of temporary cessation of production, this clause will maintain a lease for a specified period of time (often 60 or 90 days, but sometimes only 30 days), within which time operations must be commenced or production resumed. If the lease is still in its primary term, rentals may be resumed in the absence of operations or production. Beyond the primary term, production in paying quantities or continuous operations are required to maintain a lease into its "secondary term."

What constitutes "commencement" of drilling or operations that acts to preclude the payment of rentals? Individual lease provisions must be examined in the light of the state statute that prevails. "Commencement of operations" may include preparation of the surface, road building to the location, or other good faith activity. Thereafter, due diligence must be pursued in accordance with the actions of a prudent operator. "Actual drilling operations" means the well must be spud to be in compliance, with actual drilling in progress.[19]

Jurisdictions have taken contrary positions on whether a lease may be held at the end of the primary term, when no production is obtained, if a well is in the process of drilling across the expiration date of the lease. Some states permit the lessee a reasonable time to "complete" the well and obtain a market. The majority have held that if no well was "completed" by the end of the primary term, the lease would expire on the basis of lack of production.[20]

Due diligence is required in the conduct of operations. For several reasons, courts tend to rule against the lessee in interpretation of oil and gas leases. One reason is the general rule that written instruments are interpreted against the person who drafts them.[21] It is also true that most litigation involving oil and gas leases is held in the county of residence of the lessor, with the judge and jury often being his friends and neighbors. However, an operator who displays good faith and meets his obligations by substantial compliance has a fair chance that justice will be extended to him. The key to interpretation is whether a covenant created on the part of the lessee is an obligation or an option.[22] The "prudent operator" standard is continually invoked: "Whatever in the circumstances would be reasonably expected of operators of ordinary prudence, having regard to the interests of both lessor and lessee, is what is required."[23]

Offset Well Provisions

The implied covenant to protect a lease from drainage is often supplemented with an express covenant that requires the drilling of an offset well in the event a well is brought in on adjacent acreage (within a specified number of feet, ranging from 100 to 660) and is producing in paying quantities. This harks back to the "law of capture," entitling each landowner to produce all the oil he can through wells located on his own land. However, the operator may be released from drilling an offset well if he can show that such a well would not show a return on investment and operating costs.[24] The duty of the lessee to protect a lease from drainage is continuous. The lessee can become liable for payment of compensatory royalties for failure to perform under offset well provisions or implied covenants to develop.

Equipment Removal

Leases vary in the amount of time provided to the lessee to return to the premises to remove his equipment after the termination of the lease. This provision protects the lessee from the accusation of abandonment.[25]

Lesser Interest Clause

This clause is also referred to as a proportionate reduction clause or lesser estates clause. If a lease is taken that appears to cover the full interest in a tract of land, but in effect the lessor owns an undivided fractional interest, the lesser interest clause allows the lessee to reduce the rental and royalty payments in proportion to the undivided interest owned by lessor. Exceptions to reduction would be

1. Rental sum of designated amount agreed upon when the lease was taken. In this case, that portion of the lesser interest clause pertaining to delay rentals would be stricken to avoid confusion.
2. Lease description specifically states that it covers "all of lessor's undivided one-fourth interest," for example, in which case it is advisable to pay the full rental amount designated.[26]
3. Shutin gas payments based on fixed-dollar amounts.

Entirety Clause

In the event that the lessor partitions his interest subsequent to his execution of the lease, the "entirety clause" allows the lessee to develop and operate the leased premises as one entity, paying rentals and royalties in the proportion of each owner to the whole. The lessee incurs no obligations to drill offset wells on the tracts so separated out of the originally described lands.

Warranty Clause

The majority of warranty clauses are worded this way: "Lessor hereby warrants and agrees to defend the title to said land. . . ." Sophisticated landowners often insert words of limited warranty that alter the provision to read, "Lessor hereby warrants and agrees to defend the title to said land *by, through and under him only. . .* " thereby releasing themselves from having to defend a title that fails from years prior to their own acquisition.

Omission of warranty altogether might very well be unacceptable to the lessee. Many assignments of oil and gas leases, however, contain statements to the effect that warranty of title is neither guaranteed nor implied.

Subrogation

Subrogation is defined as the substitution of one party for another with reference to lawful claims or debts. A provision for subrogation generally follows the warranty clause, entitling the lessee to pay back taxes, mortgages, or other encumbrances, assuming the position of the former creditor in order to protect his own interest.

Force Majeure

Events sometimes occur that are completely out of the control of the lessee, preventing his compliance with the provisions of the lease. These occurrences are described as "force majeure" and include such events as acts of God, war, and weather. When the lessee's performance of covenants and obligations is so hindered, the force majeure clause extends the rights of the lessee by a time period equivalent to that during which his ability to act was suspended.

Signatures and Acknowledgments

The signature of the lessor should conform to the styling set out at the beginning of the lease.

It is proper for an illiterate to sign his name with his mark. Ordinarily, the signing is witnessed by one or more persons and the acknowledgment taken of the witness. Another means is to have the illiterate person designate someone to sign his name, followed by a sworn statement to the effect that the signing party was requested and authorized to sign for the lessor. The acknowledgment would be taken for the illiterate "whose name is subscribed by mark to the foregoing instrument." This would be followed by an acknowledgment of the witness, also therein stating that said witness acted in such capacity.

The form of acknowledgment used must conform to the laws of the state in which the leased premises are situated. Louisiana and certain eastern states require witnesses and acknowlegment of the witness signatures as well as of the executory party. The great majority of states provide that a notary public may take an acknowledgment. Some states require the act be done by another official or clerk of the court. In Pennsylvania, it is wise to have as subscribing witness an attorney who is licensed to practice in that state.[27] In any case, the witness should be someone other than the landman, a person having no fiduciary interest in the transaction.

There are individual acknowledgments and corporate acknowledgment forms. A party signing for another (e.g., under power of attorney or as trustee or guardian) would use the individual acknowledgment form and after the words "for the purposes and consideration therein expressed," add the words "and in the capacity therein stated."

Who has the capacity to enter into this contract? The party empowered to sign the lease. In the case of jointly owned lands, all parties must usually sign. In general, signatures are obtained from the following:

1. Married woman: joined by husband if homestead or in community property states. She may sign individually if it is her separate property, not homestead.
2. Married man: joined by wife if homestead or in community property states. He may sign individually if it is his separate property, not homestead.
3. Minor: different statutes and varying ages of majority. Out-of-state guardianship appointment may require ancillary guardianship proceedings be filed in the state where lands are situated. Usually guardian may sign, with ratification by the minor at the time he reaches majority.
4. NCM (non compos mentis): state statute determines appointment of a guardian for one with "unsound mind."
5. Trustees, executors, and administrators: under authorization of court order, will, and/or trust agreements.
6. Receivers: appointed by the court in cases of bankruptcy; also appointed for persons absented for statutory number of years who cannot be located.
7. Life tenants and remaindermen: life tenants with joinder of or ratification by remaindermen.

8. Attorney-in-fact: under valid and sufficient power of attorney. Many states require specific power to execute oil and gas lease.
9. Corporations: officer empowered by corporate resolution of board of directors.
10. Partnership: one or all partners, in accordance with statute and partnership agreement.
11. Municipalities: under procedures described by prevailing ordinance, to cover streets, parks, and other city or county properties.
12. Churches and cemeteries: trustees or other church officials so designated in that denomination.
13. Ancillary guardian: court-appointed guardian in state where property is situated. Guardian situated in "foreign" state may not be empowered to execute a lease in another state.

Implied Covenants

In addition to the expressed covenants set out in the oil and gas lease, several "implied" covenants are recognized by the courts as obligations of the lessee:

1. To drill an exploratory well, except under a lease that provides for the payment of rentals to delay drilling
2. To conduct additional development of the leased acreage "with reasonable diligence" once production in paying quantities is discovered
3. To conduct diligent and prudent operations
4. To market the product
5. To protect leased premises against drainage[28]

Riders and Special Clauses

The following are examples and suggested wordings for commonly used special provisions. Many companies and attorneys prefer specifically designed language. In the event of conflict between a printed provision in the lease and a provision in a rider, the special language of the rider prevails. All wording so added to a printed form should be initialed by the lessor or "signed for identification." The advice of legal counsel is recommended to the lessor and lessee to assure that any added provision accomplishes its intended purpose.

Exclusion of "Other Minerals"

Notwithstanding any other provisions herein contained, this lease covers and includes only oil and gas (including casinghead gas and other gaseous hydrocarbons), sulfur, and byproducts derived from oil and gas produced from the premises. All other minerals and materials are expressly reserved by Lessor, including but not exclusively, coal, lignite, uranium, and other fissionable materials.

This lease does not cover coal, iron ore, or gravel and such are excepted from this lease and reserved to Lessor.

This lease does not cover or include uranium, uranium ores, vanadium, plutonium, or other fissionable minerals or materials, and such minerals or materials are excepted from this lease and reserved to Lessor, his heirs and assigns; and it is understood and expressly provided that the term "other minerals" whenever or wherever used in this lease shall not refer to or include uranium, uranium ore, vanadium, plutonium, or other fissionable minerals or materials.

Notwithstanding anything hereinabove to the contrary, and there is excepted herefrom and reserved to the Lessor herein all uranium, fissionable materials, and all bentonite, fullers earth, and other claylike substances; it is specifically understood and agreed that this lease covers only oil, gas, sulfur, and associated liquid or liquefiable hydrocarbons, but this lease does not cover or include any other minerals, with all other minerals being reserved to the Lessor herein. Accordingly, the words "oil, gas" when used herein, shall mean oil, gas, sulfur, and associated liquid or liquefiable hydrocarbons; and the words "all other minerals" whenever used herein shall be stricken from the lease, so that such "all other minerals," as defined herein are reserved to the Lessor.

Surface/Livestock Damages

Lessee agrees to fill and level all slush pits so as to return the surface to its original condition as nearly as practical, within a reasonable length of time after the abandonment of the use of such pits. Lessee agrees to compensate Lessor for damages to growing crops, grass, livestock, and other property of Lessor located on said premises as a result of Lessee's operations on the leased premises.

Lessee agrees to observe special care during operations on the leased premises that livestock not be injured and that said livestock be protected by fencing from wandering near slush pits or equipment. Lessee agrees to protect all freshwater streams on the leased premises. In addition, Lessee agrees to fill and level all slush pits so as to return the surface to its original condition as nearly as practical, within a reasonable length of time, and to compensate Lessor for damages that do occur to growing crops, grass, livestock, and other property of

Lessor located on said leased premises as a result of Lessee's operations thereon.

Lessee shall pay Lessor for all damages to growing crops, livestock, timber, surface, buildings, water wells, personal property, and surface improvements on the leased premises resulting from Lessee's operations hereunder; and further agrees that within a reasonable time after completion or abandonment of any well drilled on the premises he shall fill and level all pits and other excavations not required for production, fencing those that must remain open, and restore the surface of the ground to as near its original condition as is practicable.

Lessee shall at all times keep closed all gates on land subject hereto and repair all damages to fences, roads, irrigation systems, and bridges inflicted by operations hereunder. Lessee shall fill all slush pits and excavations of all kinds, so as to return the surface to its original condition as nearly as practical within a reasonable length of time after the abandonment of the use of such pits and excavations. Lessee shall pay to Lessor reasonable reimbursement for damages to crops, livestock, and other property caused by operations of Lessee hereunder.

Depth Limitations

It is expressly agreed and provided that units pooled for gas, distillate, and/or condensate shall not exceed 320 acres plus a tolerance of ten percent (10%) for a depth of 9,000 feet or less below the surface of the earth, and shall not include more than 640 acres plus a tolerance of ten percent (10%) for any depth more than 9,000 feet below the surface of the earth, provided that should governmental authority having jurisdiction prescribe or permit the creation of units larger than those specified, units thereafter created may conform substantially in size with those prescribed or permitted by governmental regulations.

Anything herein to the contrary notwithstanding, it is understood that Lessor reserves unto himself, his heirs and assigns, all rights from the surface of the ground down to a depth of 1,500 feet below the surface; and Lessor, his heirs and assigns, shall have the right of ingress to and egress from the reserved premises and shall have all rights and privileges as to the reserved premises that are herein given Lessee as to the granted premises.

Anything herein to the contrary notwithstanding, it is understood and agreed that Lessor reserves unto himself, his heirs and assigns, all rights below a depth of 9,800 feet below the surface of the earth;

reserving unto Lessor, his heirs and assigns, the rights of ingress and egress to develop and produce the reserved strata, with all rights and privileges pertaining thereto.

Freestone Rider

Irrespective of the other provisions of this lease, it is agreed that should this lease be maintained in force solely by operations on or production from a pooled unit or units embracing land covered hereby and other land, such operations or production will maintain this lease in force only as to the acreage embraced in such unit or units. This lease may be maintained as to acreage not included in such unit or units in any manner provided for herein, provided that, if it be by rental payment, the rentals shall be proportionately reduced.

Pugh Clause

Irrespective of the other provisions of this lease, it is agreed that should this lease be maintained in force solely by operations on or production from a pooled unit or units embracing land covered hereby and other land, such operations or production will maintain this lease in force only as to the acreage embraced in such unit or units. This lease may be maintained as to acreage not included in such unit or units in any manner provided for herein, provided that, if it be by rental payment, the rentals shall be proportionately reduced. If at or after the end of the primary term this lease is being maintained in force as to a portion of the land solely by such operations on or production from such unit or units, Lessee may continue to maintain the lease in force as to acreage not embraced in such unit or units by rental payment for a period not in excess of two (2) years beyond the end of the primary term. Notwithstanding anything contained in this paragraph, if at any time nonunit operations should be conducted on or nonunitized production is obtained from this lease (whether or not the well is within the boundaries of a unit), then this paragraph shall not be applicable and such operations or production will serve to maintain the entirety of this lease.

It is understood that drilling operations or reworking operations on, or production from a pooled unit or units created pursuant to Para. __ hereof, shall not maintain this lease in force as to acreage covered hereby that is not included within any such unit, and this lease can only be maintained in force on acreage covered hereby that is not within any such unit by:

a. payment, during the primary term hereof and two (2) years thereafter, of delay rental pursuant to Para. 5 hereof, reduced at the rate of $2.00 per acre by the amount of acreage then covered thereby that is, at the time of such payment, included in all such units; and/or

b. shutin gas royalty payment in accordance with Para. 3 & 14 of this lease; and/or
c. drilling operations or reworking operations on or production from such acreage not included in any such unit or units, as the case may be, under the other provisions of this lease.

Upon termination of two (2) years after the primary term hereof, this lease shall terminate as to such acreage not then included in any such unit, unless Lessee is maintaining this lease in force as to such nonunitized acreage by shutin royalty payments, and/or production and/or drilling or reworking operations, as may be more particularly provided for in other provisions hereof.

Nothing herein contained, however, shall be construed as an obligation on Lessee to create any units or unit.

Option to Extend Primary Term

Lessee is hereby given the option to extend the primary term of this lease for an additional ________ from the expiration of the original primary term. This option may be exercised by Lessee at any time during the last year of the original primary term by paying the sum of ________ per net mineral acre to the parties entitled to delay rentals according to Lessee's records. This payment shall be based upon the number of net mineral acres then covered by this lease, and all of the provisions of this lease relating to the payment of delay rentals shall apply equally to this payment including, but not limited to, the provisions regarding changes in ownership. If, at the time this payment is made, various parties are entitled to specific amounts of the delay rental according to Lessee's records, this payment may be divided between said parties and paid in the same proportion. Should this option be exercised as herein provided, it shall be considered for all purposes as though this lease originally provided for a primary term of ________, except that no delay rental payment shall be required to maintain this lease for the first ________ months of the extended term. In the event this lease is being maintained by any provision hereof at the expiration of the original primary term, Lessee shall have a period of thirty (30) days from the date this lease ceases to be so maintained within which to exercise this option.

Nonuse of Surface

Lessee agrees that he will conduct no operations on the surface of said land. It is further agreed and understood that Lessee shall have the right to drill and operate directional wells through and under said land, irrespective of the bottom hole locations of said wells. To this end, Lessor hereby grants to Lessee a subsurface easement for all purposes associated with such directional wells.

Neither Lessees herein, nor their heirs, successors, or assigns shall have any right of ingress or egress in or on the property hereby leased, or any right to the use of any part of the surface of said land for exploration for, or development, production, or handling of any such oil and gas except to the extent necessary for the creation and formation of a pool or unit in accordance with the applicable regulations of the ________ Commission of the State of ________.

Limitation of Shutin Well Payments

It is understood that the "shutin gas well" provisions of Para. __ of this lease shall be a recurring right, limited to an aggregate or cumulative total of three (3) years.

It is expressly agreed and provided that this lease cannot and shall not be maintained after the expiration of the primary term solely by payment of the shutin well royalties in accordance with Para. __ for a longer period than two (2) consecutive years or for shorter periods at various intervals not to exceed in the aggregate two (2) years in all; provided, however, this right to maintain this lease after the expiration of the primary term for a two-year period shall be a recurring right and may be exercised from time to time whenever the Lessee finds it necessary to shut in such well.

Accelerated Delay Rentals

It is expressly understood that notwithstanding anything contained herein to the contrary, the amount to be paid by Lessee as provided for in Para. __ hereof called "rental" shall be increased during the last two (2) years of the primary term hereof to the sum of $________, the payment of which sum on or before the third and fourth anniversary dates of this lease shall give Lessee the right to further defer the commencement of drilling operations during the remainder of the primary term as provided for in Para. __.

All Premises to Be Unitized

Notwithstanding any of the other provisions herein to the contrary, it is understood and agreed between the parties hereto that should any of the leased premises covered hereby be included in a unit for gas or oil, then the entire premises covered by this lease shall be included in any such unit.

Other Special Provisions

Various agreements may be made with the lessor regarding accelerated delay rentals, accelerated royalties, compensatory royalties in the event of drainage, or provisions for minimum royalties. The nature of these particular stipulations requires legal counsel for appropriate language.

Bonus Payments

Fee leases are most frequently paid by bank draft drawn on the lessee's bank (or to the account of his lease broker, if prearranged). The lessor takes his original executed lease to his bank with the draft, placing both items with the collection department for further handling.

The lessor's bank sends the items to the collection bank, which is responsible for collecting the monies due before releasing the executed instrument to the lessee. This procedure protects both parties in that the lease would be returned to the lessor if the draft were not timely paid. The lessee may return the lease and draft unpaid if, upon inspection of the instrument, he finds the lease is improperly executed or a further title check has revealed a failure or discrepancy in ownership.

Drafts are drawn for a specified number of days for payment, from the date of arrival at the collecting bank. For instance, a "sight" draft is due upon arrival. However, 15- and 30-day drafts are most common in order to allow time for confirmation of title information. Due dates range from 5 to 45 days, as a rule, but they may be for longer periods. When the draft is paid, funds are forwarded by the collecting bank to the account of the lessor at his bank of origination.

Bonus payments may be made by check or by cash if a proper receipt is given. Payments may be given upon execution of the lease instrument. Deferred bonus payments can be arranged between the parties to accommodate the lessor. In the event of a deferred bonus, a letter between the parties should set out the exact manner of payment and should state that the lease is to be considered in full force and effect from its effective date.

Option to Lease

Under certain circumstances, a landowner might grant an option to lease his minerals. The lessee might request an option to allow himself time to put together his financing or well deal.

A portion of the bonus money may be advanced as prepayment, with the balance due on exercise of option. An election not to exercise the option would forfeit the initial payment. Another contract might provide for an advance payment over and above any future expected bonus sum. Terms are negotiable, depending entirely on how each party deems them to be to his advantage.

Top Leasing

A "top lease" is a lease taken on a mineral interest that is under a valid lease presently in effect.

A lessee may "top lease" his own lease that has not yet expired. Frequently, a provision is added to his top lease to terminate the existing lease and make the top lease effective immediately. This accomplishes a dual purpose: First, any conflict between the terms of the two leases is removed in favor of the new lease. Second, a cloud on the title is prevented by terminating the old lease before it might be thrown into a secondary term under a continuous operations clause or by pooling.

The top lease may be taken by another party lessee on lands currently under lease to a prior lessee. The proper wording of effective dates has already been discussed in this chapter. A top lease may contain a clarifying provision that it shall become effective only upon expiration of the lease that presently exists. The top lease becomes a conveyance of the lessor's reversionary interest. Otherwise, a top lease can be held void under the rule of perpetuities: It may never become effective because the current lease might not expire within the perpetuity period. Legal counsel should approve any special language to be added to a lease that will constitute a top lease.

Recordation

Contract law provides that an effective contract must include delivery. Recordation in the public records of the county or parish where the lands are situated constitutes constructive notice of delivery. So-called "dresser drawer" instruments and contracts are discovered from time to time which were not recorded until many years after the date of execution.

The general rule prevails that the earliest instrument of record is the effective instrument.

Leasing Procedure Summary for Lessor

The prospective lessee may leave a lease and draft for the bonus sum with the lessor. If the lessor has read and understands the terms of the lease and is in accord with the negotiation, the lessor may proceed in the following manner to consummate the transaction:

1. Properly execute the lease, and have signatures appropriately acknowledged.
2. Fill in name of depository bank on face of lease, including address of bank and your account number.
3. Place Tax ID number or Social Security number below signature on lease.
4. Endorse the draft, and include Social Security number.
5. Place draft and fully executed lease at the collection department of your bank for handling. They will be forwarded to the collecting bank and the money returned to lessor's account when the draft is paid.

Federal Leases

The Federal Mineral Leasing Act of 1920 reserves to the United States government a fee interest in all original domestic public lands. The Mineral Leasing Act for Acquired Lands covers properties acquired by the federal government from other owners. The latter includes most of our military lands. National parks, monuments, naval and national petroleum reserves, and certain wilderness areas are excluded from leasing.

Oil and gas leasing procedures for federal lands are governed by the secretary of the interior. Lands are subject to classification or withdrawal from use, at the secretary's discretion, for the public good. Leasing is governed under exacting restrictions and continuing supervision from the Code of Federal Regulations (CFR), Title 43, as amended. Anyone interested in filing for a federal lease is advised to request a copy of CFR 43 and all amendments to date from the nearest office of the Bureau of Land Management. This office will also provide copies of necessary forms and procedures for filing. Federal lands are continually affected by policies and regulations adopted under environmental laws. At this writing, the BLM works jointly with the Minerals Management Service (MMS) to administer public lands for leasing.

Simultaneous Oil and Gas Lease Filings

The simultaneous drawing system is a lottery. One may apply only on lands that have been previously leased. Copies of the list of available lands for each drawing may be obtained from the nearest

BLM office by written request, with fee paid in advance. Parties interested in monthly lists and results of drawings are allowed to place a deposit with the appropriate BLM office. Fees are charged against the deposit as a declining balance account.

Lists of available lands are posted on the first of each odd-numbered month, with offers accepted during the following 15 days. Automated computed forms are processed in two parts: Form #3122-6 Part A and Form #3112-6a Part B.

Part A provides identity information for the applicant and need be filed only one time. Part B is the parcel application. Forms may be obtained for a nominal fee from the Superintendent of Documents, U.S. Government Printing Office, Washington, D.C. 20402. Small orders may be acquired at no cost from BLM state offices. Filing fees are $75 per parcel, payable by postal money order or certified check.

Drawing results on parcels applied for are mailed automatically. Entire result lists may be ordered from state BLM offices.

General Requirements Concerning Public Domain

A simultaneous or noncompetitive lease provides for a 10-year primary term. Delay rentals are $1 per acre, payable in advance. Assignments of interest are valid only after approval of application for transfer of interest, accompanied by a $25 application fee. Royalty transfers are by approval of application form, submitted with a $10 fee.

Each operator is required to post a general lease and drilling bond or an operator's bond, in the minimum sum of $10,000. Active operators may function under a $25,000 statewide bond or a $150,000 nationwide bond.

Leases may be acquired only by citizens of the United Sates or by companies and corporations of such citizens that are organized under the laws of the United States.

Full disclosure is required of the applicant of the names of all parties owning or holding any beneficial interest in the offer to lease, including both oral understandings and written agreements. A statement of interest and a copy of any written agreement shall be filed with the BLM no more than 15 days after the application to lease. The applicant should own 50% or more of the operating rights, or the lease offer shall be subject to rejection.

Agreements creating in excess of an aggregate 5% overriding royalty (including production payments) on oil are deemed in violation, unless the agreement suspends excess overrides or payments when average production per well is 15 bo/d or less. This limitation of overriding royalty is not applicable to gas production.

Certain acreage limitations per parcel also apply: no offer may exceed 10,240 acres, no offer may be for fewer than 640 acres; and an offer must lie within a six-square-mile area. Miscellaneous exceptions to these requirements are described in 43 CFR.

Successful applicants under simultaneous drawings are notified upon receipt of the lease agreement and request for first year's rental. Executed lease and rental payment within 30 days constitutes acceptance.

Maximum holdings by one person or entity are limited to 246,080 acres in any one state except Alaska, where the limit is 300,000 acres in the northern leasing district and 300,000 acres in the southern leasing district.

Wildlife refuge lands may be leased only with consent of the secretary of the interior and the concurrence of the U.S. Fish and Wildlife Service. Game reserves are subject to like limitations. The U.S. Corps of Engineers oversees water facilities. Special forms are available from individual agencies or from the area BLM office for rights-of-way or temporary-use grants.

Public Lands Not Previously Leased

Official BLM tract books and records are open for search and inspection. Filings may be made "over the counter" for parcels found to be available for leasing. Such filings require five signed copies of BLM form 3110-1, a $75 fee, and the first year's rental, which may be submitted by mail or in person.

Federal Lease Forms

Bureau of Land Management Form 3110-1 serves as an "Offer to Lease and Lease for Oil and Gas" under the simultaneous filing system for noncompetititve public domain leases. Special instructions on this form call for:

1. Name and address of offeror
2. Description of lands
3. Enclosure of $75 filing fee plus first year's rental @ $1 per acre or fraction thereof (two checks)
4. Statement of U.S. citizenship and age
5. Cumulative federal acreage under lease by offeror not to exceed legal limits

Full instructions for filing are enumerated on the reverse side of the offer form. A principal change in regulations not indicated on Form 3110-1 is an increase in the filing fee from $10 to $75 per bid.

The signature of the offeror on the form constitutes his acceptance should he be awarded the lease in the simultaneous drawing. The offer cannot be withdrawn once a lease for the subject acreage has been signed on behalf of the United States. The purpose of two checks (certified checks or money orders) in payment of fee and rental allows the BLM to retain the filing fee on all bids (successful or unsuccessful) and to return the rental checks to unsuccessful bidders.

Lease terms are also printed on the reverse side of the offer form. Significant provisions of the lease include:

1. Ten-year primary term "and as long thereafter."
2. Lessee must file and maintain the required bonds.
3. To drill and produce any wells necessary to protect the premises from drainage (or, with permission of the director, to pay a compensatory royalty to lessor).
4. Rentals of $1 per acre (and any portion thereof) annually for lands outside a known geologic structure. If any of the leased premises are included in such KGS, 30 days notice will be given to lessee to increase the rentals for those acres to $2 per acre or portion thereof.
5. After discovery of production, to pay a minimum royalty of $1 per acre or portion thereof annually at the end of each lease year. If paid royalty from production amounts to less than $1 per acre, the difference must be made up. Should only a portion of the lands be pooled, minimum royalties apply only to participating acreage. The secretary likewise may waive any rental or royalty payments if it is deemed by him to be in the interest of encouraging further production and development.

   Pooling or unitization is only with consent of the secretary of the interior.
6. Royalty to lessor is one-eighth (12.5%). However, the secretary of the interior is empowered to raise the royalty at his discretion (before execution of the lease).

   The secretary may also set minimum prices for the products on which the royalty is based.
7. Failure to pay timely rental causes automatic termination of nonproducing acreage.
8. Copies of any sales contracts for oil, gas, or other products shall be furnished to the appropriate BLM officer.
9. Well data, production information, and reports shall be furnished upon request.
10. Taxes shall be paid by lessee, as assessed.

11. Lessee shall comply with terms of equal opportunity laws.
12. Assignments or transfers of interest shall be upon approval of lessor. Lessee is required to file copies of operating agreements, subleases, or other transfers (recordable or nonrecordable) within 90 days of consummation.
13. Protection and restoration of surface. Lessor may prescribe steps necessary to restore the surface.
14. Overriding royalties are limited to 5%.

Restrictions and reservations set out to the lessor are essentially:

1. The right to surface easements and rights-of-way
2. The right to sell or dispose of surface, if it does not interfere with lessee's operations
3. The right to take and extract helium
4. The right to take royalties in kind or value, to purchase casing, and to operate water wells

At this writing, the BLM is in a stage of transition and merger, having taken over the responsibilities under federal minerals programs from both the U.S. Geological Survey (USGS) and the Minerals Management Service. In December 1982 the USGS conservation division merged with MMS. On that date the USGS relinquished its functions dealing with federal minerals. The USGS remains a division of the Department of the Interior, dealing with maps, water facilities, and various geological services. The intent of the merger into BLM is to create one office with one line of authority with whom lessees may deal.

Acquired Lands

Applications to lease acquired lands are made by means of Form 3110-3, which is very similar to that for simultaneous filings. This form must be filed in seven original counterparts.

Primary terms for leases on acquired lands are stipulated at 10 years. Royalties to the lessor are one-eighth. Overriding royalties are limited to 5%. The majority of the provisions coincide with those of Form 3110-1. Variations deal with undivided fractional interests that may be owned by the United States.

Competitive Oil and Gas Lease Bids

Competitive sales for leases on KGS lands are held periodically after appropriate advertising by the BLM. Sealed bids are accepted on Form 3000-2. The high bid is determined by bonus and royalty offered.

The sealed bid must be submitted with a cashier's check, money order, or bank draft covering a minimum of one fifth of the bonus offered. Full disclosure of all parties in interest is voluntary, but the bid may be rejected if the information requested is not provided.

Competitive leasc forms are issued to the successful bidder for a primary term of five years "and as long thereafter."

State Leases

Leases covering state-owned minerals are regulated and administered by the appropriate state land office or land commissioner in accordance with the individual state's statutes. The following states are selected for discussion as representative examples.

California

The State Lands Commission of California processes applications and conducts competitive bidding on state-owned minerals. Provisions vary. Terms may be negotiable. Bids are awarded on a sliding-scale royalty, ranging from 16⅔% to 50%, as specified on the invitation to bid.

Awards are based on four methods of bidding, according to that in the best interest of the state:

1. Highest cash bonus, with royalty set by the commission
2. Highest multiplying factor offered on royalty
3. Highest flat-rate royalty
4. Highest percentage of net profits

Negotiated flat-rate royalties may be as high as 40%. Net profits offered range from 25% to 81%, in addition to royalty.

The primary term of the lease is designated as 20 years "and as long thereafter" as production or operations continue. However, the drilling term for well commencement is set at two to three years from the date of the lease.

Rental rates vary from $1 to $50 per acre, depending on private activity in the area.

Each lease is subject to environmental assessment. The cost of environmental documentation is paid by the nominator of the tract, who is reimbursed by the successful bidder if he be another party. The successful bidder may also be subject to the costs of an environmental impact report if it is deemed necessary.

Leasing is subject to the Public Resources Code, Div. 6, Part 2, Chap. 3, as amended. The filing fee is $25 plus a $100 nonrefundable deposit.

A lease may be assigned in part or in whole upon approval of the commission. The state reserves the right to sell or dispose of the surface or to grant any surface easement. Bond is required of the lessee. Copies of contracts, logs, drilling records, and well histories must be furnished to the commission. These documents are held in confidence.

Units including state-owned lands may be formed only with state approval. Drilling requirements and general accounting procedures for net profits are attached to the lease as exhibits.

Regulations and forms may be acquired from the State Land Commission, 245 West Broadway, Suite 425, Long Beach, California 90802.

Colorado

Oil and gas leases are granted in Colorado on a competitive basis, generally by a sale held on the third Wednesday of each month. Filing fees are $12.75 for the first 160 acres, plus $1.50 for each additional 160 acres or portion thereof. Leases are awarded to the highest bidder of bonus per acre.

Primary terms are for five years with an option to renew for five years by written application. Rentals of $1 per acre double during the extended term. Royalties to the lessor are one-eighth.

Monthly production reports are required. Pooling and unitization are by consent of the lessor. Shutin royalty may be waived by the state until the 11th year of the lease, at which time the payments shall be $2.00 per acre ($320 minimum), payable for a maximum term of five years.

Assignments are by written consent of the lessor, on a 40-acre minimum.

Colorado leases are administered by the State Board of Land Commissioners, 620 Centennial Building, 1313 Sherman Street, Denver, Colorado 80203.

New Mexico

New Mexico segregates its public lands into two "districts": (1) restricted, comprising proven oil and gas areas, and (2) unrestricted, all lands outside restricted areas.

Regularly advertised sales are listed the third Tuesday in each month. Applications to lease must be accompanied by a $10 filing fee and a check for bonus and first year's rental, presented by sealed bid. Leases are awarded to the highest bonus bidder.

Primary terms for unrestricted lands are five years, with an option to renew for a secondary five-year term. Restricted lands are leased for five years only "and as long thereafter."

Royalties are one-eighth on unrestricted lands and one-eighth to one-sixth on restricted acreage. No lease shall cover more than two sections, but no limit is set on thc number of leases one may acquire.

The lease forms for each type of acreage are parallel, the major exception relating to terms and royalty. Rentals provide for $1 per acre ($6 minimum) during the first five years and double during the secondary term.

The lessor reserves the option to take in kind or to purchase at the prevailing market price in the *area.*

Operations continuing beyond the end of the primary term shall be approved by the state on written application, prior to the termination date. A saving clause provides for 60 days with no cessation of operations for more than 20 days.

Copies of contracts and agreements pertaining to the lease shall be furnished to the commissioner or filed in the county deed records.

Formal application is necessary for the formation of any unit. Assignments are accepted on certain conditions:

1. A prescribed form from the State Land Office must be filed in triplicate with a $25 fee.
2. The assignee's entire interest must be shown.
3. Minimum acreage covered is one legal subdivision.
4. Acreage from only one lease may be in an assignment.

The assignment divests the assignor of his full obligation and responsibility.

New Mexico state lands are administered by the Commissioner of Public Lands, P.O. Box 1148, Santa Fe, New Mexico, 87504.

Louisiana

Three procedures are necessary to lease state-owned minerals in Louisiana:

1. Written submittal to the Secretary, Department of Natural Resources, requesting advertisement of the tract. Application includes detailed description of lands, plat, and $300 deposit made by certified check or money order.
2. Advertisement of tracts for lease are placed by the state in the Official State Journal and in the Official Parish Journal.
3. Bids are to be submitted in separate sealed envelopes with tract number designated on the front of each. The bids are then placed in a mailing envelope and sent to the Office of Mineral Resources by registered mail. All bids are opened publicly on specified day of sale.

Minimum royalty is set in advance for each tract and is noted in the sale advertisement. Annual rentals shall not be less than one-half the amount of the bonus offered. Lease terms shall not exceed three years. The state reserves the right to reject any bid deemed not in the best interest of the state. Bids are considered on the basis of bonus and royalty offered.

Lands administered by state agencies are leased by similar procedures. The agency may lease its own lands or may ask the State Mineral Board to handle the leasing. State agencies include levee, drainage, road, and school districts; municipalities and state institutions. Lands acquired by the state under tax sales require that applications to lease include names of persons and dates of adjudication.

The Louisiana state lease form (rev. 1981) for inland properties is a multipage instrument of stringent requirements. Among its major provisions are:

1. Express covenants to drill offset wells and to protect lands from drainage
2. Compensatory royalty provisions
3. Lessor reserves option to take royalty in kind; value of royalties based on price prevailing in the field
4. Formation of unit requires application to the Louisiana commissioner of conservation with applicable plats and geological information
5. Ninety-day saving clause
6. Semiannual shutin payments based on $25 per lease acre ($500 minimum)
7. Assignment with approval of state
8. Two-year Pugh clause with deferred development payments on the "outside acreage"
9. Environment codes and protection of water facilities
10. Penalty of $100 per day for failure to release following 30 days after date of termination
11. Special rider for sale of oil or gas to intrastate markets

Applications and information may be obtained from the State Mineral Board, Office of Mineral Resources, P.O. Box 2827, Baton Rouge, Louisiana 70821.

Oklahoma

Bidding on state lands or state school lands in Oklahoma first requires a written request for advertisement of the tract, accompanied by $100 advance deposit on costs of advertising. Notice is posted in the county newspaper for 15 days prior to the sale.

Each tract shall be bid upon separately in sealed envelopes placed inside the mailing envelope. At least 25% of the amount bid must be remitted at the time the bid is offered. The minimum bid recognized is $5 per acre. Consideration of conditions other than highest bonus are advertised under a "Special Oil and Gas Lease Sale."

A lease owner may appoint another interest owner as operator by filing the appropriate form. Seismic operations require a fee and special permit. Assignments of interest in a lease may be on any recognized industry form but are subject to the approval of the commissioner. Assignments subject to other agreements shall be accompanied by a copy of such agreements. Assignments of overriding royalty do not require commission approval.

Upon termination of the lease, the lessee has 90 days to restore the surface and remove equipment, subject to a penalty of $100 per month for noncompliance.

Lease Form CL00-101 (rev. March 1, 1982) for the State of Oklahoma provides for a three-sixteenth lessor royalty and a three-year primary term. Rentals are $1 per acre annually. Shutin payments shall be equal to annual rentals. At the end of the primary term, the lease terminates as to all nonproducing formations.

Operations will not be deemed to have commenced until the lessee files a notice of intention to drill at least five days prior to commencement.

Surface owners shall have free use of gas for domestic purposes. (This provision is not acceptable to many operators because of the hazard and liability involved in using household gas to which no odor has been added.)

Oklahoma state lands and leases are governed by the Commissioner of the Land Office, Fourth Floor, Jim Thorpe Building, State Capitol, Oklahoma City, Oklahoma 73105.

Texas

Texas lease forms and procedures are published for five classifications of state-owned lands.

1. Relinquishment Act (Mineral Classified) Lands. These lands are subject to the Relinquishment Act of 1919 and include state-owned lands on which the surface was sold and minerals reserved to the state between the years 1895 and 1931. Sales transacted during that time reserving no minerals are not subject to the Relinquishment Act.

Green v. Robinson, 117 Tex. 516, 8 S.W. 2d 655 (1928), effectively established the owner of the soil as the agent for the state for leasing. The surface owner and the state split equally all monies received under the terms of the lease, including bonus monies. Thus a surface owner cannot sell the land and reserve any interest. Only the present surface owner may negotiate and execute oil and gas leases. He may not use an agent or attorney-in-fact.

Negotiated leases (Form 5367, rev. 1981) are by private transaction with the surface owner. The lease must cite the actual consideration. It is subject to ratification by the General Land Office (GLO), which reserves the right to reject the offer.

General provisions include a primary term not to exceed five years. A minimum rental of 10 cents per acre is payable annually to the state, regardless of production. Shutin royalty is double the annual rental, but a minimum of $1,200 annually. No top leases or extensions are allowed.

2. Free Royalty Lands. Lands sold after August 21, 1931, subject to the Sales Act of 1931, reserve to the state a one-sixteenth nonparticipating royalty.

3. University of Texas System Lands. The University of Texas Board for Lease has its own lease form and sale procedures. It sets sale dates and terms. Leases are by public auction on a competitive bidding basis, after due notice of the sale.

Special requirements limit the use of the surface. A payment is required along with application for maintaining a lease by drilling across the end of the primary term. A technical staff makes recommendations to the board regarding pooling decisions.

4. State Fee Leases. Riverbeds, channels, tidelands, islands, lakes, and bays are under this classification, along with lands situated in the Gulf of Mexico.

Provisions include three-year primary terms, with the exception of riverbeds, which carry two-year terms. Rentals are $5 per acre, royalties to lessors one-fourth, and minimum royalty equal to annual delay rental. The shutin provision is notoriously difficult to interpret and apply. Annual shutin payments of $1,200 will hold a lease, but shutin status is interrelated with other producing wells situated within 1,000 feet of the leased premises and the payment of compensatory royalties.

Drilling across the end of the primary term requires advance application for extension and payment of $3,000 for leases comprising up to 640 acres, $6,000 for leases comprising more than 640 acres. The application is for a period of 30 days and must be renewed as needed.

Leasing of riverbeds and channels is restricted to those within two miles of production, with exceptions.

5. State Agencies, Boards, Institutions, and School Lands. Each agency has its own board for lease. A special lease form is similar to that for fee leases. Procedures for parks and wildlife lands, the Department of Corrections, Texas Highway Department, mental health institutions, and lands belonging to the Alabama Coushatta Indians require nomination of the tract by application to the GLO. Sales are conducted on the first Tuesday of the month, but not necessarily every month. Sealed bids are to be accompanied by two checks to cover the entire bonus and 1% of the bonus as a fee. Tie bids are rejected, and a new sale is scheduled. High bonus is awarded the lease, except when royalty bids are specified (generally on lands offsetting production).

The School Land Board lease provides for three-year terms, the others for five-year terms. A minimum royalty of one-fourth is designated on school lands. A Pugh clause covers "outside acreage."

Pooling is granted by special approval of the land commissioner. Application for pooling should include maps of the leases, a unit outline, and geology that justifies the action. The lessee must show that pooling is necessary and in the best interest of the state on the basis of lease expiration date, size of unit, and economics. Pooling will likely be limited to the product (oil or gas) and to the producing formation(s). A GLO pooling agreement form is used.

Leases covering lands belonging to Texas A & M University, Texas A & I University, and Texas Tech University are handled separately but under similar regulations.

The state of Texas provides loans to veterans for the purchase of private lands. The Veterans' Land Board requires a special lease form to be used for leasing lands subject to veteran loans. This lease provides for equal payments to be made to the Veterans' Land Board and to the lessor until the loan is paid off. The lease must be ratified by the chairman of the Veterans' Land Board to be effective. Special provisions include protection of freshwater strata, sale of natural gas to intrastate market, and free use of gas by the lessor for household purposes. Shutin provisions are onerous and similar to the wording in state fee leases. Payments for shutin are twice the annual rental ($300 minimum) for each well.

Copies of applications, regulations, and forms may be obtained from the General Land Office, Austin, Texas 78701, or from the Veterans' Land Board in the same city.

Wyoming

Public lands in Wyoming that are under the jurisdiction of the state include its fee lands, school lands, farm loan acreage, institutional lands, and properties acquired through escheat. The last of these are subject to a stipulation that the state assumes no liability or warranty of title.

Leases are offered through competitive bidding only. Specified sale dates are announced, with terms and conditions of the sale included in the sale notice. A list of the available tracts for bidding is published. To nominate a particular tract for the sale, application on the proper form must be filed, accompanied by a $25 filing fee, a minimum bonus payment of $1.00/ac. and an advance rental payment of $1.00/ac. Tender must be by cashier's check or money order. Nominations should be received by the Commissioner at least 60 days prior to sale date. Each nomination shall not exceed 640 acres. Nominated tracts are not identified to the public until after the sale is concluded. Bidding is oral, with each lease awarded to the highest bidder. Personal checks are accepted.

A Wyoming lease form (eff. Febr. 1, 1982) provides for a five-year primary term, one-sixth lessor royalty, and $1 per acre rentals, payable annually in advance. An interesting rule in Wyoming allows late payment of rental upon notice of default from the state and a penalty of 50 cents per acre payable with the late rental, within 30 days of notice. Rental payments increase to $2 per acre after production is found, with such payments credited to royalty for that year.

Top leasing is not permitted, but the term of a lease may be extended one year upon written request. The extension is conditional upon commencement of drilling operations within the extended term and diligent completion of the well. A performance bond is required to cover liquidated damages that shall be due to the state upon default.

The state board is authorized to require a unit plan if it is deemed necessary to protect the state's interest. Application of the lessee to form a unit requires submittal of two copies of the preliminary unit agreement, two copies of plat, and geological information to substantiate the request.

If only a portion of the leased acreage is included in a unit, the balance is considered an "assignment"; that is, it segregates the remaining acreage as if it were two separate leases, creating new obligations on the nonunitized land. An assignment of undivided interest in the lease is acceptable if no more than two parties hold interest. An excess of two parties requires that one be designated as agent or attorney-in-fact for all.

Cost-free interests, such as overrides and production payments, are subject to certain conditions:

1. If not exceeding 5%, they may be reduced in the same proportion that the state agrees to reduce its royalty if the state finds such a move necessary to promote full development.
2. If they are in excess of 5%, the assignee must agree that one-half of the interest will be paid to the state of Wyoming. The assignee must also agree that the interest may be cancelled by the board if it finds that the burdens interfere with reasonable development.

Surrender of leases upon which operations have been conducted is subject to an inspection report by the state oil and gas supervisor.

Information from the state of Wyoming is available through the Commissioner of Public Lands, 2424 Pioneer Avenue, Cheyenne, Wyoming 82002.

Indian Leases

Regulations governing the leasing of Indian lands that belong to organized tribes may be found in the Code of Federal Regulations (CFR), Title 25. The lands are controlled by the Bureau of Indian Affairs.

Indian tribes may lease their lands with the approval of the secretary of the interior.

The Bureau of Indian Affairs (BIA), following authorization from the appropriate tribal council, publishes a notice 30 days prior to the lease sale. Bids are accepted and leases awarded on a competitive basis to the highest bonus bidder. Each sale specifies the terms to apply. Bids must be accompanied by a cashier's check or money order for no less than 25% of the bonus amount offered. Within 30 days after notice of acceptance, the successful bidder must remit the balance of the bonus plus the first year's rental in advance. He also pays his portion of the advertising costs of the lease sale. Bids may be rejected by the secretary of the interior when he feels that it is in the best interest of the tribal council to do so.

The lessee must furnish bond in the designated sum on the basis of the number of acres leased. A lessee may acquire more than one lease, but no single lease shall be comprised of more than 2,560 acres of land for oil and gas. Each leased area is to be a "reasonably compact" body of acreage.

The primary term of an oil and gas lease on Indian lands shall not exceed 10 years, and such terms are more frequently for five years. Rentals are $1.25 per acre, payable annually in advance. Most leases specify a one-eighth royalty. The secretary of the interior may fix a higher royalty, prior to advertising the tract, at his discretion. Gas prices at the time of production must be calculated at a value equal to the highest price received for the product in the field where leased premises are located.

Every lease, permit, assignment, or sublease requires a filing fee of $10. Each assignment must be approved by the secretary of the interior, and each new owner must post a performance bond. Agreements creating overriding royalties and production payments are not considered interests in the lease that require the secretary's consent. If such conveyances or instruments are included in the terms of a conveyance to be filed for record, a copy should be furnished to the superintendent of the area BIA office. All owners of undivided interest must join in an application for cancellation or surrender of a lease.

Regulations and restrictions are stringently enforced. Noncompliance by the lessee may void the lease. The provisions of a tribal charter, bylaw, or constitution may supercede the regulations established under CFR 25. For example, Crow Nation leases are renewable for additional 10-year terms. Osage lands may be extended 5 years. The Wind River Tribe may execute 20-year term leases, renewable for 10-year periods.[29]

Joint-venture legislation is under consideration by Congress to restructure present Indian lease agreements.[30] Proposed changes would encourage participation by the tribal councils in exploration and production agreements and service contracts. The secretary of the interior would monitor joint ventures as to the economic benefits for the tribes and as to environmental impact.

Indian land leases are increasingly subject to negotiation rather than to fixed terms. Allotted lands, in particular, are including agreements in leases for exploration and development, net proceeds payments, and working interest options.[31] Navajo and Southern Ute tribes currently require 25% royalties plus carried interests or net profits.

The extended waiting periods for leases to be issued and delivered create serious problems for the lessee. One case enumerated a lease that was issued but not delivered for over a year, with substantial delay rentals due and payable without the lease in hand.[32] A minimum delay is generally 60 days. A tribal council may execute a lease on its lands and return it promptly, but this is the exception rather than the rule.

A lease taken on allotted lands requires the lessee to locate and secure the signature of each person known to have any interest in the tract. The BIA or tribal agency will furnish the prospective lessee a list of owners and last known addresses. Funds are not necessary to the transactions, but it is the lessee's responsibility to persuade each owner to execute the lease if and when he is found. The absence of one signature prevents the lease from becoming effective. The lease is awarded on a 30-day basis, with 30-day extensions if all signatures are not acquired within the first period. In the meantime, the tribe retains the bonus monies. North Dakota and Montana are notorious areas for problems of finding and dealing with allottees. The best advice to one desiring to lease allotted lands is that prior to bidding he should ask for a list of persons from whom signatures would be required on that tract. If the list is imposing, another tract might prove to be less formidable and perhaps more advantageous in the long run.

References

[1]Leslie Moses, *The AAPL Guide for Landmen,* rev. ed. by Milton L. Brand (Tulsa: Kraftbilt, 1980), p. 13.

[2]Ibid.

[3]John S. Lowe, "Anatomy of an Oil & Gas Lease," *Oil & Gas Law for Attorneys* (Oklahoma City: The Institute for Energy Development, 1982), p. 139.

[4]Moses, p. 15.

[5]AAPL Educational Programs, Inc., "The Oil, Gas & Mineral Lease," *Advanced Land Practices* (Houston: American Association of Petroleum Landmen, November 1982).

[6]Alfred A. Ring and Nelson L. North, *Real Estate Principles and Practices,* 6th ed. (Englewood Cliffs, N.J.: Prentice-Hall, 1967), pp. 120–121.

[7]Earl A. Brown, *The Law of Oil and Gas Leases* (New York: Matthew Bender, 1958), par. 3.01, pp. 21–22.

[8]Ibid., p. 22.

[9]Lewis G. Mosburg, Jr., *Landman's Handbook on Petroleum Land Titles* (Fort Worth: Institutes for Energy Development, 1976), p. 59.

[10]Brown, par. 4.03, pp. 55–57.

[11]Ibid., p. 53.

[12]Richard W. Hemingway, *The Law of Oil & Gas* (St. Paul, Minn.: West Publishing Co., 1971), p. 333.

[13]Lowe, p. 141.

[14]*Southwestern Law Journal,* vol. 36, no. 1, Annual Survey of Texas Law, Mineral Resources.

[15]Howard R. Williams and Charles J. Meyers, *Oil and Gas Law,* abr. ed. (New York: Matthew Bender, 1975), par. 632, p. 350.

[16]Ibid., par. 606.2, pp. 314–315.

[17]Lowe, p. 143.

[18]Black, Henry Campbell, *Black's Law Dictionary,* p. 1510.

[19]Ibid.

[20]Williams and Meyers, par. 617, p. 334.

[21]Lowe, p. 146.

[22]Ibid., p. 142.

[23]Brown, par. 16.02, p. 296.

[24]Ibid., par, 16.02, p. 315.

[25]Lowe, p. 147.

[26]Brown, par. 14.04, pp. 246–247.

[27]Edwin T. Stitt, "Eastern Overthrust Belt," *Landman* (June 1982), p. 31.

[28]Brown, par. 16.02, pp. 269–274.

[29]Daniel H. Israel, "Current Legal Issues Affecting Oil & Gas Development on Indian Reservations," *Landman* (November 1982), p. 50.

[30]Ibid., p. 53.

[31]Wendell E. Haley, CPL, Transcontinent Oil Company, Denver, Colorado, interview, January 4, 1983.

[32]Ibid.

CHAPTER SIX

Curing Titles

Title Opinions and Abstracts

A title opinion is prepared by an attorney after his examination of the records or of an abstract. An abstract is composed of one or more bound books comprised of copies of all instruments (up to the date indicated) that are of record in the county where the land is situated and that affect the particular parcel being examined. Abstracts are prepared by an abstract or title company that is usually located in the same town as the county seat. The company's records are gathered from county, district court, and tax records.

California is one state in which the landman may rely on a title policy in lieu of a title opinion. At the request of the landman, a title company prepares a preliminary report that enumerates encumbrances and exceptions in the title. Curative is accomplished for each case in accordance with the directions of the landman's company attorney. A title policy may then be issued in the amount of money requested to cover the risk.[1]

To order an abstract, one must furnish the abstractor with the following information:

1. Legal description of land to be covered.
2. Period of time to be included, e.g., sovereignty to March 31, 1982, 8:00 A.M. (This would be known as a base abstract.)
3. If it is to be a complemental or supplemental abstract, the number and dates of the abstract it will supplement (add on to) or complement (add to data included in another abstract for an overlapping time period). For example, a supplemental abstract might be ordered to bring up-to-date Abstract 16789, prepared by Gofer Abstract Company, that covers the period from September 23, 1961, 7:00 A.M. to June 1, 1963, 8:00 A.M. The supplement ordered would bring the latter date to the present time. A complemental abstract is ordered similarly, e.g., complemental to Abstract 18899, prepared by Gofer

Abstract Company, covering ______ described lands, from February 23, 1943, 7:00 A.M. to date. In this case, the complemental abstract would cover the lands specified, whereas the one in hand might cover only a part of the lands to be examined and possibly other lands.

Time and money are saved if base or other abstracts can be borrowed from the landowner. Abstractors frequently have a backlog of orders. It is not unknown to wait a year for the preparation of an abstract, although two to three months is more common. Meanwhile, time is running on the leases, and rentals are falling due.

There may be several hundred instruments of a certain type on record that do not affect the mineral, royalty, or leasehold title (e.g., unit ratifications). If the company attorney prefers to do so, he may elect to have these instruments deleted from the abstract when it is ordered. Not all abstractors are willing to take the responsibility for omitting instruments from the abstract.

If there is no local title company or abstractor, if the abstractor is too busy to get the abstract out in time, or if the cost is greater than the company wishes to bear, the examining attorney might be asked to make a "stand up" opinion, that is, to go to the courthouse and prepare his opinion directly from personal examination of the records. This is time consuming and is the least preferred method for a busy attorney. However, such an opinion may be expedited if the landman makes a run sheet—a list of *all* instruments of record, in chronological order according to file date. Included in this list would be instruments from the county clerk, probate, deed of trust, county and district courts, and lis pendens (suits pending).

Three general types of title opinions serve varying purposes. The drillsite title opinion covers the drillsite lease(s) and its provisions. It is prepared prior to the drilling of the well. The attorney must determine that the company owning the working interest holds valid leases from the proper persons who own all of the mineral interest under the tract. He ascertains which curative must be finished prior to the drilling of the well so that no cloud on the title would prove a challenge in court if a good well is brought in. He calls attention to onerous or ambiguous lease provisions.

An original title opinion may be requested and prepared. This opinion is more extensive. It establishes who the persons are who own good title to the entire premises and includes an analysis of the mineral, royalty, overriding royalty, leasehold working interest, and surface ownership.

The last type of title opinion to be requested is a division order opinion that sets out proper division of royalty payments in the event of production.

All of the opinions may have requirements made by the examining attorney in order for the curative processes to clear the record title. Requirements may be requests for affidavits, subordination agreements, copies of probates filed in other counties, tax certificates, or copies of marriage or death records. The landman's job is to comply with the requests of the examining attorney as fully and as completely as possible. The landman should first read carefully, word for word, the title opinion that he plans to cure or send someone else to the field to cure. He should attempt to satisfy the requirements precisely in order to avoid having a supplemental opinion issued that states "Requirement not satisfied."

Certain title imperfections may be waived at the business risk of the company. Other requirements are headed "Advisory," meaning the company may make a business decision to waive that requirement without risking a defective title. If a long period of time elapses between the date the original title opinion is issued and the receipt of the curative, the attorney may ask for a new supplemental abstract to bring the records up-to-date for his examination.

A conscientious landman heads off requirements and curative at the time he acquires the lease by being certain it is properly executed and acknowledged, by avoiding riders with ambiguous language, and by submitting the necessary copies of probates or affidavits when the lease is brought in.

Curatives

Certified copies of curative instruments may be specifically requested for filing in the county where the land is situated. Sometimes certified copies are the only type of copy furnished by the public office. A certified copy costs more per page than a plain photocopy. It has the certification of the clerk that the instrument copy furnished represents a true and correct copy of the instrument that has been filed for record. Only certified copies are recordable for filing elsewhere.

Public offices often allow the use of their photocopy machines on an honor system. Other offices require that one of their employees do the photocopy work. If the landman plans to use a particular county office on a regular basis, he may be able to set up an account to pay for copies on a weekly or monthly basis. A sight draft system might be used, whereby each certified copy is paid for with a sight draft on the checking account at the bank of the person requesting the copy. This type of account must be set up ahead of time, with a notice to one's bank to honor the sight drafts. (Many banks charge a fee for each draft.)

Affidavits

Affidavits may be prepared from printed forms or drafted and typed on plain legal bond paper. If the typed form is used, the instrument is titled "Affidavit of ________," the state and county names are typed at the far left, as on the printed form, and the wording begins with the customary statement:

> Before me, the undersigned authority, on this day personally appeared ________, of lawful age, who being first duly sworn, deposes and says:

This is followed by a statement in the words of the affiant. An affidavit requires the signature of the affiant and this sworn statement (jurat) following his signature:

> Sworn to and subscribed before me, this ____ day of ______, 198____

The jurat is taken by a notary public or other authorized official. Below the jurat is an acknowledgment, stating that the person who signed is the person he says he is, and that he executed the instrument for the purposes therein expressed. Unless the affiant's signature is acknowledged, the affidavit is not a recordable instrument. Many affidavits require the signature of a witness, with the acknowledgment of the witness taken by a notary public or other official empowered to do so. Certain states do not allow the recording of affidavits.[2]

An affidavit of heirship details the full name of the deceased, when and where he died, his age at death, his relationship to the affiant, whether he left a will and if so whether it was probated, whether an executor or administrator was appointed for the estate, marital status at time of death, and information as to previous marrages (if none, so state). Also included are the names of all surviving children, with their dates of birth, current addresses, and names of spouses; and the names of all children who may have predeceased the person about whom the affidavit is taken, when they died, and names of their spouses and their surviving children.

In the event that the deceased left no will, an affidavit of heirship will be required for any child or surviving spouse who has since died. If the deceased had no children, the names of his surviving spouse, parents, brothers, and sisters would be listed, along with their addresses and the names of their spouses. It is imperative that the landman be familiar with the laws of descent and distribution in the state where the land is situated and also in the state of domicile of the deceased, which may determine his heirship. It may be necessary to ask an attorney to research this information in the event of "foreign" (out-of-state) estates.

The affidavit might include information about the land in question, whether it was a homestead, whether either the deceased or his spouse ever lived on the land, and whether it was considered to be separate or community property.

An affidavit of use, possession, and occupancy or an affidavit of adverse possession is a detailed history of the use of the land described, as far back as the affiant can recall. The affidavit would state the relationship of the affiant to the facts (neighbor, relative, etc.) whether the lands were fenced, where the fences were located, whether there were gates or cattle guards, whether these are still standing, and whether the fences would turn livestock. Information is taken regarding all structures or wells on the premises, who built them and whether they are still standing or in use, or who tore them down. Other data would include timber use, crops planted, tenants who may have farmed the land, animals raised; who occupied the premises and on what dates (years); and who the neighbors were at the time. An affidavit of adverse possession would state information regarding any boundary disputes or occupation of the land by cotenants or adverse possessors. Public roads crossing the land, easements, ditches, or prominent landmarks would be set forth and described. Production history would be included, with the affiant's knowledge of any wells drilled on the lands, by whom, and whether the wells produced or have been plugged and abandoned. A statement should be taken regarding the location of any churches, cemeteries, schools, or railroad crossings, that have been or are now situated on the premises, or the absence of these.

An affidavit of identity is used to establish that a person who has used more than one name of record, or differing initials, is the same person. This affidavit would state, for example, that James J. Smith is one and the same person as J. John Smith, also known as (aka) J. J. Smith, or "Jay" Smith. It might establish that a married woman,

Margaret W. Fortress, is the wife of Horace Fortress, that she was born Margaret Ann Stonewall, was first married to I. M. Weakley, was later known as Marge Weakly, a widow. All possible combinations of names by which a person has been known and could be confused should be included.

Affidavits of nonproduction are taken when a release of an old oil and gas lease cannot be acquired because the company is dissolved or the parties deceased. An affidavit of production history might be needed to establish the cotenancy of the operator or working interest owner with the mineral owners against any adverse possessor.

Other types of affidavits include homestead or nonhomestead, marital, divorce, and other status in question.

When the landmen feels there is a doubt regarding the validity of the information given in an affidavit (due to lack of memory or lack of knowledge on the part of affiant), supporting affidavits should be taken from other persons familiar with the circumstances. Two or three affidavits of use and possession are frequently required.

Personal Inspections

The astute landman will make a trip to the site he is leasing, at his first opportunity. He can knowledgeably discuss the lands and visualize the problems as he deals with the landowners. When a personal inspection of the leased premises is required by the examining attorney, the landman will make a trip to the site and prepare a plat to scale, detailing the physical aspects of the land: outline of boundaries; location of structures and their condition; types of fences and gates; power lines, pipelines, roads, wellsites (including water wells); crops, trees, livestock. The plat submitted by the landman must be neat, must always indicate North, and must be identified by lease name, county, state, and prospect name. A written report accompanies the plat.

Ratifications

A ratification affirms that the person who executes it agrees to the terms of another instrument cited therein by recording reference and date. Ratifications may be taken from remaindermen to affirm that they are in accord with and join in the terms of an oil and gas lease that has been executed by the owner of the life estate under which they hold. A ratification may be taken from a spouse whose signature was omitted from a lease on joint property. The first spouse who signed the lease would also join in the ratification.[3] The ratifying party may not become entitled to a portion of the bonus, rentals, or royalties under the terms of the lease unless the ratification so provides. Ratifications for purposes of pooling are taken from owners of nonparticipating royalty, overriding royalty, or a nonparticipating mineral interest. Should some mishap occur regarding the timely payment of a rental, a ratification by the lessor might establish that the lease continues in full force and effect.

Frequently, if a questionable interest is outstanding or clouds the title, a "protection lease" is acquired from the unestablished owner. Protection leases create title difficulties in themselves. A "cleaner" method of acquiring signatures for a questionable interest might be by means of a ratification.[4] The language of the ratification is important regarding whether the ratifying party will share in bonus and rental monies and if so to what extent.

Subordinations of Lien

A mortgage recorded after the recordation of an oil and gas lease is subject to the lease, and no subordination is necessary.[5] If the mortgage were recorded prior to the lease, a subordination of lien might be required in the title opinion to subordinate the lien to the lease in the event of foreclosure, so that the validity of the lease would not be affected. A subordination of lien is a printed form that differs from state to state.

The landman prepares a formal letter of request to accompany the subordination form he has completed with the pertinent information. The letter is directed to the president or loan officer at the appropriate bank or mortgage company. If it can be presented and personally requested, the landman has the opportunity to explain the need for the instrument. Certain institutions will not execute a subordination.

Other subordinations might be requested for deeds of trust or vendor's liens or by surface mining lessees.

Tenant's Consent Agreements or Damage Releases

To prevent the surface owner or tenant from making future claims against the operator of a producing well, a damage release may be acquired to settle anticipated losses in advance. Even though the surface estate is frequently subordinate to the mineral estate, the damage release might clear the way for peaceable entry onto the premises and establish good relations. A tenant's release would cover a monetary settlement for crop or livestock damages due to operations or for the construction of facilities or pipelines. Common wording is

> for all damages of every kind and nature to the land, both surface and subsurface, and such livestock, crops, grass, fences, timber, and all other improvements located on the land resulting from *(name of operator)*'s drilling of the *(name of well)*, located in *(county and state)*, and by your acceptance, Operator is hereby released from all damages due to you as a result of the above-described operations.

If the landowner will not make a settlement prior to operations, an estimated settlement that offers protection for both parties should be reached if possible.

Tax Certificates

The tax assessor-collector in the county where the land is situated will furnish a tax certificate for a small fee. The certificate states that taxes have been paid or are unpaid, by whom, and for which years. Tax certificates are acquired from the tax office for ad valorem taxes, from the school tax office for school taxes, and from a junior college district where applicable. Tract descriptions in the tax offices do not always match descriptions of the lands in deed records. The tax office may have the tract subdivided by surface ownership, with taxes rendered to several persons in addition to the lessor or other than the lessor. Minerals are usually assessed only after production, but this matter varies from county to county and state to state. Certificates as to all parties owning an interest in the lands in question would be taken for curative purposes. This is also an appropriate time to determine whether any tax liens are outstanding (local, state, or federal) or whether current tax sales have been held covering any of the described lands.

Quitclaim Deeds or Disclaimers of Interest

A divorced spouse, adverse claimant, or squatter might be asked to execute a quitclaim deed or disclaimer of interest to remove a cloud on a title. Such a cloud might be removed by one who has executed an instrument of record but who owns no interest in the property.

Boundary Agreements

If the location of a boundary poses any dispute between neighboring owners, a survey plat might be acquired or a boundary agreement taken. A boundary agreement is executed by all concerned parties; it states the exact location of the line between two lands and the owners' agreement to this decision.

Releases

Unless a state statute requires a release or the instrument contains a provision requiring a release, the lessee is under no firm obligation to provide one. Most companies are cooperative and will file a release on request. Many leases contain a provision allowing for partial release, particularly of nonproducing acreage. Partial releases are often filed by holders of deeds of trust when a portion of the mortgage is paid. In Louisiana, any and all owners of a beneficial interest in a lease are required to sign a release and place it of record.

Surface leases, mining leases, liens, or judgments against the land need to be released upon termination. In certain cases, an action to quiet title (trespass to try title, or TTT suit) might be the most appropriate means to establish clear title, even though court processes are expensive and time consuming. The landowner is the most likely party to pay for such action. However, the lessee might decide it is to his benefit to institute the suit on behalf of the landowner.

Probates

Often the will of a deceased owner in the chain of title has been filed for probate in a county or state other than the one in which the land is situated. Probates are filed in the state and county of domicile. The landman must determine where these matters are recorded and obtain copies for examination by an attorney. Certified copies of the application, will, order to probate, and inheritance tax receipts are generally sufficient. Probate proceedings include many instruments that are not needed and are expensive to reproduce; among these are inventories, bonds, letters testamentary, proof of will, etc. A will may have a trust agreement incorporated into the document. The landman should read these items carefully and summarize his findings regarding the devise of the property in question when he makes his report.

Miscellaneous

Other curative instruments that are often required are copies of trust agreements, corporate resolutions, and powers of attorney—all of which establish the right of one person to sign in the capacity of another. Trust agreements may be requested from banks or individuals, but they are not always furnished. Corporate resolutions are obtained from the corporation authorizing the officer to act on its behalf. A power of attorney should be of record somewhere, but it may not yet be recorded in the county where the land is situated. One can obtain a certified copy from the person acting as agent and attorney-in-fact by way of the principal for whom he acts or from public records.

Correction deeds are filed for the purpose of amending or clarifying a previously recorded instrument. If a correction deed that conflicts with the previous deed of record is filed, the terms of the correction deed will prevail.

Stipulations of interest are commonly prepared to set out the varying undivided interests of several parties holding under one assignment or deed. Rental division orders indicate the distribution of rentals if two or more lessors have executed the same lease instrument and wish to have their rentals distributed individually.

Sources of Information

Where does the landman locate information on persons who are long since dead or when he does not know where or if they are living? Suppose the title opinion states that a person is a stranger to the title, and the requirement is to find out who he is and how he fits into the chain of ownership. Where does one begin? Following is a list of sources of information:

Tax office records
City directories
Telephone directories
Postmaster interviews
Local abstractor's records
County or district Clerk's records
Marriage-birth-probate records
Attorney's and doctor's files
Neighbors' recollections
Cemetery records and headstones
Funeral home records
Old newspapers (copies in libraries)
Genealogies (in state and local libraries)
Census records (county, state, and federal)
State land office correspondence
State comptroller's records
Retirement home files
Hospital records
Secretary of state's records of defunct or current corporations (in state offices)
Department of Defense, Washington, D.C.

Curing titles is like fitting together a jigsaw puzzle. There are many pieces. With persistence, they may all fit.

Run Sheets

One word of caution to the landman preparing a run sheet for the examining attorney. The attorney will rely on the run sheet. It must be inclusive, but it should not contain extraneous matters that do not pertain to the title. A familiar name in the chain of title may have numerous suits against him for nonpayment of small monetary debts, or he may be in judicial proceedings with regard to other lands. The landman must read these instruments to determine if they should be included. He should include any bankruptcy proceeding, which may indicate a distribution of assets to creditors, the appointment of a receiver, or a possible trustee's deed. He should look for and include in chronological order, according to filing date (but also include instrument date), all conveyances, deeds of trust, vendor's liens, affidavits, powers of attorney, unit agreements, dissolution of units, tenancy agreements, surface leases, subordinations, divorce proceedings, probates, lis pendens, boundary agreements, and/or litigation of any nature affecting ownership of the lands. An obvious gap in the title is designated by a line of dots, as used in the index of an abstract.

Each page of the run sheet identifies the lands searched and is numbered. The date of search is important. It is printed on the first page to indicate the last date on which the records were checked. The daily file register in the clerk's office brings the run sheet up-to-date.

It is wise to check the county records and the abstractor's records if time permits. Now and then the abstractor's chain will omit an instrument. Miscellaneous cards of the abstract records supply information regarding other instruments that may apply to persons in the chain of title.

Title Curative Reports

As each curative instrument is acquired, it should be stapled with all pages pertaining to its question. An identifying label or small typed note stapled to the upper left-hand corner (e.g., "Requirement 6: William T. Martin, Agt. & Atty.-in-fact for Jane Doe"; or "Dow Jones, dec'd. 1-1-81, Probate") makes these items readily identifiable.

The curative report is a formal letter addressed to the person for whom the landman works or to the attorney upon such instruction.

It will state in the opening paragraph that "Mr. Bob Bullett prepared his Original Title Opinion on _______ lands, by date of _______" and that the curative enclosed is in accordance with the requirements as set out in that opinion, with page numbers as referred.

It is recommended that each requirement be set out and described separately: "Requirement 3, p. 4." The landman may briefly describe the curative furnished and how it accomplishes the objective. He is advised to keep the curative letter brief and to the point. If a requirement is not satisfied, he should state why it is not and whether it will be furnished at a later date. "Respectfully submitted" is an appropriate closing statement to the letter.

Curative materials are attached to the letter with a strong clip. A photocopy of the title opinion being cured is included along with a plat of the lands. The easier the job of the attorney, the more successful the efforts of the landman.

References

[1]Leslie Moses, "From Lease to Release," *The AAPL Guide for Landmen,* rev. ed. by Milton L. Brand (Tulsa: Kraftbilt, 1980), p. 95.

[2]Lewis G. Mosburg, Jr., *Landman's Handbook on Petroleum Land Titles* (Fort Worth: Institutes for Energy Development, 1976), p. 43.

[3]Ibid., p. 119.

[4]Ibid., p. 120.

[5]Ibid., p. 66.

CHAPTER SEVEN

Pooling and Unitization

The terms "pooling" and "unitization" are similar in purpose but separately defined.

Pooling refers to the combining or communitizing of two or more small tracts that are, in themselves, too small for a well to be permitted on each tract, or that by pooling would more efficiently be developed. "The purpose of pooling is to prevent physical and economic waste" that occurs by drainage or by the drilling of unnecessary wells.[1] Pooling allows an operator to comply with spacing regulations for one-well units. It prevents the drilling of multiple wells when one would serve to deplete the reservoir. Pooling protects the correlative, or mutual, rights of the mineral and royalty owners to a common source of production. Owners under the three-, five-, and ten-acre tracts are afforded the opportunity of a well when pooled with additional acreage to form a "drilling" or "proration" unit (one of legally designated size). Oil units are usually 40 to 80 acres in surface area; gas wells require 160 to 640 ± acres. Conservation statutes that regulate spacing seek a goal of rational and controlled development to provide maximum production at minimum expense. A well is generally required to be located a certain number of feet from the lease line to prevent drainage of an offsetting tract. Exceptions to the spacing rules allow a well to be permitted on a tract that is shown to be in the danger of losing its pro rata share of production.

Royalty owners in a pooled unit receive proceeds from sales on an apportionment basis: each according to the ratio of his surface acreage to the entire pooled unit times the lessor royalty provided in his lease. Pooling clauses often specify the method by which the royalty will be paid to the lessor. Generally, royalty is apportioned and paid according to "tract factor":

The lessor's undivided interest in the tract times the ratio of that tract to the acreage in the unit times the royalty fraction provided in the lease times gross unit production.

A pooling of sorts is accomplished by means of "communitization" or the "community lease." This may be accomplished with a single lease that describes several tracts (usually adjacent, but not always), with all mineral owners under the several tracts executing the same lease (or a multiple counterpart). Royalties are apportioned in proportion to the acreage participation of each owner.

The pooling of one tract with another under the pooling provision is not to be confused with the powers given to the lessee in an "entirety clause." An entirety clause allows the lessee to treat the entire producing leased premises under one lease *as though* it were pooled, in the event that one or more parcels are subdivided out of the lands subsequent to the execution of the lease. Each owner shares on the apportionment basis, according to surface participation. Without an entirety clause in the lease, each partitioned interest would be treated on a nonapportionment basis: only the owner of the block on which the well was situated would receive proceeds from production. "Courts have held that the entirety clause is a covenant running with the land which restricts the power of parties subsequent to alienate the mineral estate free of the entirety clause."[2]

Apportionment of royalties is neither applicable nor practical in the instance where a mineral owner severs his interest horizontally, subsequent to the execution of an oil and gas lease on the entire premises. For example, the grantee who receives all interest below 6,000 feet would consequently receive all royalties due and payable to that tract for any well that should produce below the 6,000-foot depth.

The rule of capture, as it applies to correlative rights, does not intend that all owners under a common source share in any production therefrom. It implies that the right of each owner is to recover or retrieve as much oil and gas as he can from his own lands. If this fair opportunity to capture is denied by spacing statutes for conservation purposes, his correlative rights are served by pooling. Pooling essentially creates one leased entity for operational purposes.

A question of the correlative rights of owners of a common reservoir leads to pooling for the common good. Owners are protected by law against negligence and waste of the operator by intent.[3] Waste may apply to the product itself or to economic waste.

Pooling provisions which state that the lessee shall not be obligated to drill offset wells could, in some circumstances, create drainage of the nonunitized (outside) acreage that offsets the unit. Protection of those lands may be afforded by the implied covenant to develop, depending upon the circumstances and geological information available. The doctrine of correlative rights initiated the first pooling. The migratory nature of oil and gas was recognized, and multiple owners of interest over a common reservoir were afforded equal opportunities to share in the production. Each owner has legal rights, but none may act to injure the source of supply or take any undue portion to himself.[4]

Unitization is the joinder of two or more tracts to form a larger unit, either for a field or for a portion of a field. Its purpose is to treat the reservoir as a common source for secondary recovery operations. Pooling and unitization apply to the consolidation of separate lands. Whether the unit is large or small, authorization comes from the lease or through a statutory requirement or by adoption by the lessor of a unit agreement amending the terms of his lease to allow unitization.

The primary purpose of the pooling provision in the lease is to allow the operator to unilaterally form a unit when and if it becomes necessary or advantageous, without the delay of recontacting all lessors for permission. But the pooling authority in the lease does not always bind the owner of a nonparticipating or overriding royalty. Permission to pool these interests is obtained by requesting the royalty owner to ratify the lease covering his tract, thereby agreeing to the pooling clause within.

Voluntary Pooling

Pooling alone does not maintain a lease. Production, or resumption of operations, is the factor that maintains the lease in the absence of rental or shutin payment or other lease provision. Termination of production or operations causes a termination of the unit. At that point, each lease exists or expires by its own terms.

The standard pooling provision within a lease authorizes the operator/lessee to form units under the rules and regulations of the applicable regulatory agency in the particular state. A "unit" generally refers to an area that can be drained by one well. The clause may allow the lease to be pooled in part or in whole with any other lands in the vicinity. Pooling may cover all minerals, or it may be restricted to oil and gas and related hydrocarbons within the phraseology of the clause. Maximum sizes for oil and gas units are described, with a tolerance factor of 10% being the norm. It is common in Colorado and in many other states for the size of the unit to be that which is *prescribed* by the governing authority. A suit in Texas styled Jones v. Killingsworth (1965) brought forth a decision from the Texas Supreme Court that wording which allows a unit size prescribed by the governing authority shall be strictly interpreted. Since the Jones v. Killingsworth decision, many pooling clauses include the words "prescribed or permitted," to render the phrase more flexible.

A lessee who exercises his pooling power must act in good faith. Fundamentally, pooling constitutes an intent to develop. The pooling process or formation of a unit should probably be accomplished before production commences. Pooled tracts are usually contiguous by the nature of the common source, but this does not necessarily constitute a requirement.

Commencement of production or of operations from any land included in the unit constitutes production or operations from all lands included in the unit insofar as perpetuating the leases is concerned. A pooling clause that refers solely to unit production may not serve to maintain a lease in the event of a shutin well or of reworking operations to restore production if the well is situated on another unit tract. Specific words bind the lessee, whose authority extends only to the rights expressly granted. The question arises as to whether "commencement of operations" on any portion of the unitized area before production is found would maintain another lease in the unit nearing the end of its primary term if the provision for extension is "as long thereafter as *production* is obtained." The pooling clause might provide that *operations* on any portion of the pooled unit would hold the lease, which eliminates the problem. In the event of a conflict between the pooling provision and any other provision of the lease, it is generally held that the pooling clause prevails.[5]

Consideration must be given to the "dry hole" clause, as it relates to pooled acreage. The pooling provision may not specifically mention that the drilling of a dry hole on lands pooled or unitized with the leased premises serves to hold the lease, even though the well is situated on another tract. John S. Lowe asserts that unit operations, even for a dry hole, should serve to satisfy the provisions of a dry hole clause in any of the pooled leases. The same is true for "cessation of operations," "resumption of operations," and "continuous operations" provisions.[6] Each lease, one would hope, would contain a saving clause to furnish the operator a specific or reasonable number of days between cessations of production or operations.

The Louisiana Mineral Code, art. 33, discusses the interruption of prescription as it relates to unitized lands. Unit operations on lands other than those burdened by a mineral servitude (severed mineral interests) will constitute effective interruption of prescription, but only as to the lands included within the unit boundaries, and only if the operations are intended for producing from the unitized sand or sands. The "outside" acreage would not be affected by such operations, nor would operations to produce from mineral strata other than those for which the unit was formed. A shutin well on unitized acreage interrupts prescription in the same manner, as of the date of the order creating the unit (Louisiana Mineral Code, art. 35).

Unless language is inserted to the contrary, any portion of the leased acreage pooled into a unit serves to hold by production or operations all of the premises described in that lease. Common restrictive paragraphs added to leases to prevent a partially pooled tract from holding a lease in its entirety are the so-called Freestone rider and the Pugh clause. The Freestone rider was first used in and is named for Freestone County, Texas. It usually provides that rentals shall be paid on "outside acreage." The outside acreage expires at the end of the primary term in the absence of compliance with other lease provisions to keep it alive.

The Pugh clause (designed by the late Lawrence G. Pugh, Sr., attorney from Crowley, Louisiana) segregates nonunitized acreage from pooled or unitized lands, allowing unit production to hold only the acreage included in the unit (whether it be oil or gas). A decision under Rist v. Westhoma Oil Co., 385 P.2d 791 (Okla. Sup. Ct. 1963), construes the Pugh clause to apply only to a vertical division of the leased acreage. An application of the Pugh clause to horizontal division would present difficult problems in complying with delay rental and other lease payments and provisions. Outside acreage may be maintained as otherwise provided in the lease.

The basic difference between the Freestone rider and the Pugh clause is the additional number of years that the Pugh clause provides for nonunitized acreage to be held by rental payments *beyond* the primary term. The rentals are proportionately reduced to the number of acres so maintained, as under the Freestone rider. The number of years for extension beyond the primary term is a negotiable factor and ranges from one to five years. The purpose of the extension is to furnish the lessee additional development time once the initial well has been drilled. The lessor must decide when he executes the lease whether it is to his advantage to allow the lessee this privilege.

Certain leases provide for the continuation of rental payments on pooled acreage after production is found. Whether such payments are considered a portion of or an advance on royalty may affect whether failure to pay would result in a forfeiture of the lease.

May a nonconsenting co-owner prevent the pooling of a lease for the other consenting owners? As a rule, no. If the lessee holds a substantial leasehold interest that has agreed to pool, he may proceed. The royalties of the nonconsenting party may be held in suspense according to most statutes. Acceptance of such royalties by the former nonconsenting party would constitute an agreement to pool.

The pooling clause might be explicit as to whether the unit designation (or declaration) may be filed before or after the well is drilled or commenced. Each lease must be examined to determine how and when pooling may be accomplished. Certain leases contain language which states that a well on another tract, drilled before the unit is formed, does not constitute a "unit well." Another lease may allow pooling even after production is "found." Pooling accomplished before or after production, "when in lessee's judgment it is necessary or advisable to do so," expresses authority for the lessee to pool productive acreage with nonproducing acreage. Royalty owners under the productive acreage might challenge the propriety of the action, arguing against the lessee's exercise of good faith.

Each unit filed should designate an effective date for the unit agreement and should set out the terms by which the agreement shall terminate. The date of effectiveness is pertinent to the terms of each lease and to the proper application of royalties. It is not sufficient that a lessee of several adjoining tracts decides that they shall be pooled. An instrument of pooling or a unit designation should be prepared in a form for recording. If two or more lessees exercise authority to pool, a unit agreement is written to state the commitment of each to unit operations. Many unit designations or agreements are effective upon recording in the county or parish records. Proration units, designed to satisfy spacing regulations, are also on file with the state governing body. Often a plat accompanies the unit declaration instrument. At times, the acreage is described as an exhibit attached, enumerating by legal description the leases to be unitized.

May the lessee alter or enlarge the size of a unit after it has been formally designated and filed? The pooling clause (or the conservation commission that has ordered the unit) must permit such action. Rarely may a unit be reduced in configuration. Generally, it may be amended to include additional leases that lie within the original surface boundaries. Enlargement of a unit would have to be in conformation with the unit sizes established by the governing authority.

Exercise of pooling authority may occur many times unless the lease contains conditions that limit the number of units which may be formed or the number of times the lessee may exercise pooling authority. Units may overlap in surface outlines. Unitized or pooled leases may cover all depths, or a unit may be restricted in its designation to particular producing horizons. A depth limitation may specify "from the surface of the earth down to one hundred feet (100′) below the base of the ________ formation" or "one hundred feet (100′) below the deepest depth drilled." By such wording, the remaining depths are not included or considered to be unitized. The unit designation may describe certain producing horizons only, e.g., from "5,682 feet through 5,706 feet."

A unit well must produce in accordance with the purpose for which it was formed. An oil unit must produce oil; a gas unit must produce substantially gas. Courts have supported the contention that a unit well which ceases to produce its stipulated product no longer qualifies as the unit well for which it was declared. Effectively, it becomes a dry hole for unit purposes.[7] If that unit is not to expire, other measures must be taken to maintain it (e.g., drilling a new unit well or commencing operations to recomplete the existing well in another zone that will comply with terms of the unit declaration).

Production from a properly authorized and pooled unit will extend and perpetuate a terminable deed covering any of said lands for mineral or royalty interest conveyed for a certain number of years "and as long thereafter as production is obtained." Uncertainty arises over whether a term mineral or term royalty interest may be perpetuated by shutin payments after the expiration of the term and whether it perpetuates under brief cessations of production. If "production in paying quantities" is necessary to extend a term interest, certainly legal counsel should be sought on the occurrence of cessation of production after the termination date, even if the unit continues to be effective.

Imes v. Globe Oil and Refining Company, 184 Oka. 79, 84 P.2d 1106 (1938), held that even though a lease pooling clause empowered the lessee to pool "with any other leases," the lessee had an implied obligation not to pool with valueless and known-to-be-worthless lands.[8] The acreage pooled should properly lie within the productive limits of the reservoir.

Pooling provisions vary from lease to lease and from state to state. A few of the usual checkpoints to consider are:

1. Should pooling be allowed when, in the lessee's judgment, it is right to do so? Or is pooling better left to the discretion of a governing authority?
2. Is pooling restricted to specific purposes?
3. Is pooling of all substances allowed?
4. May the premises be pooled with noncontiguous lands?
5. Does the lease provide that if any of leased lands are pooled, all must be pooled?
6. Is language included for units to be sized as may be "prescribed or permitted" by the governing authority?
7. How shall royalty be allocated on the pooled acreage?
8. Are rental or shutin payments due on the pooled acreage as well as nonpooled acreage?
9. May unit boundaries be altered?
10. Must the unit declaration be filed for record to be effective; is an instrument of termination required in the event production ceases or for other reason the unit is no longer in effect?
11. Are the designated maximum unit sizes amplified by a "tolerance" factor (usually 10%)?
12. How many times may the lessee exercise the authority to pool?
13. If pooling is allowed only for conservation purposes, does the lessee have an implied "good faith" duty to pool?[9]

A unit agreement or designation may provide the terms under which the unit will terminate. If an attempt to establish production fails, the unit may terminate. A gas well that ceases to produce gas will not maintain an established gas unit; the same applies to an oil well for a designated oil unit. Production may cease altogether, with no further operations planned because the reservoir is depleted. One unit may terminate when it is superseded by formation of a statutory unit. The unit would cease to exist if, by particular circumstances, all leases but one expired (or if all expired, of course). Many leases require that a dissolution or termination of unit instrument be filed for record when the unit is no longer in legal force and effect.

Compulsory Pooling

Compulsory, or forced, pooling laws have evolved in various states for the preservation of the correlative rights of small parcel owners, as well as for purposes of conservation. Kansas is the only actively producing state at this time that has not enacted a statute for forced pooling.

The basic premises upon which the laws have been passed are:

1. Excessive drilling tends to dissipate natural reservoirs and produces economic and physical waste.
2. Uniform well spacing allows for conservation and equitable production.
3. The small tract owners may participate in production from a reservoir attributable to his lands without an unnecessary well being drilled on a small parcel and without the confiscation of his minerals by adjoining landowners.[10]

Statutes for compulsory pooling were first enacted in Oklahoma and New Mexico. Oklahoma declared that a lessee who owned a majority interest in a unit was required to include other lessees and to share in the cost of development as well as to share in production. Other states followed suit. Colorado statute provides that under an order for compulsory pooling, an owner who refuses to agree upon the terms for drilling and subsequently refuses to pay his proportionate share of costs shall have such costs paid out of his share of production as reimbursement to the operator. Other states concur. However, reimbursement to the operator comes only from proceeds from production under that specified well. If the well is dry or ceases to produce, the operator has no recourse against the nonconsenting party. New Mexico allows a charge for the risk taken by the operator, a charge not to exceed 50% of the well costs. Pennsylvania has also enacted a nonconsent penalty, not to exceed 100% of the costs.

Procedures vary from state to state in the methods used to form force-pooled units. Generally, a petition is filed with the proper regulatory agency. Information is furnished to the agency regarding the location and type of well proposed, size of unit, reasons supporting the application to pool, and how costs shall be borne. Notices are sent to interested parties by the agency or commission, inviting them to appear and testify or comment at a hearing. The hearing is held on the designated date. If all is in order and the commission is in accord with the plans, an order is issued for the establishment of the unit. If the working interest owners have not agreed to the allocation of costs, the order may set the terms and conditions to be followed.[11]

In the state of Oklahoma, the Corporation Commission regulates the oil and gas activity. The commission is empowered to enforce compulsory pooling and to establish and prescribe the size of a drilling unit. Units that are created by the commission are based on a "common source of supply," depending upon the geological considerations. Drilling units may be established whether or not they are drilled, and the units remain in effect until the commission acts to dissolve them. The Corporation Commission does not maintain complete records on all units that have been established. It is recommended that one desiring to check a title use an Oklahoma City firm named Oil-Law Records Corporation, which has a full computer-retrieval system of information.[12]

During the hearings for forced pooling, the commission will establish a fair current market value for a lease. The mineral owner who chooses not to participate in pooling may find himself subject to compulsory pooling law, accepting the market value set by the commission. Diligent efforts are made to contact all mineral owners in the subject area in order to encourage them to attend the hearings. Owners are urged to pool voluntarily.[13]

The legislature of the state of Texas vested powers of conservation in the Railroad Commission of Texas in 1919, giving rise to the first well spacing regulation in the United States, which is known as "Rule 37." Rule 37 requires that wells be located certain distances from lease boundaries and certain distances from other wells on the same lease. Exceptions arose through the years, each designed to prevent waste or "confiscation." The confiscation exception allowed each tract—no matter how small—one well, either on that tract or on land with which it was pooled. No tract may be intentionally isolated by pooling around it. On the other hand, no lands may be intentionally subdivided to provide closer well spacing. The effect of Rule 37 has been to allow oil units of a minimum size of 40 acres. Prior to the designation of proration rules (control of production by the governing authority), more than 25,000 wells were drilled in an East Texas field covering five counties. Fire hazard was prevalent, and the price of oil dropped below 10 cents per barrel in 1931.[14] Litigations abounded, but the United States Supreme Court upheld proration and the right of the Railroad Commission of Texas to set the rules (Railroad Commission v. Rowan and Nichols Oil Co., 310 U.S. 573, 60 S.Ct. 1021, 1940).[15] Policies of the Railroad Commission were altered after World War II.

When the drilling of a well on a small tract of land was not feasible and the owner of the tract was refused entry into a pooled unit, the owner was allowed to receive his fair share of the production after drilling and operating costs were recouped by the operator.[16]

The Texas Mineral Interest Pooling Act, applicable to reservoirs discovered after March 8, 1961, seeks to encourage "compulsory voluntary pooling." There must be owners of separate interests within a proration unit who have not agreed to pool, but the key is that they must have been made a fair and reasonable offer to do so. The nonconsenting owner who is force-pooled may either pay his share of the drilling costs in advance or his proportionate share of the costs will be deducted from production to cover drilling and completion expenses.[17] "The Supreme Court has held that the royalty owner may not invoke the act," but only the working interest owner.[18]

Other states that have compulsory pooling and unitization statutes include Arkansas, Louisiana, Iowa, Nebraska, Kentucky, Nevada, North Dakota, Illinois, Florida, Georgia, New Mexico, Alabama, Kansas, and Washington. The laws of Iowa, Nebraska, North Dakota, Illinois, Kentucky, and New Mexico appear to support the rights of the royalty owner to petition for compulsory pooling.[19] Owners of nonparticipating interests such as overriding royalties and production payments are generally not afforded the opportunity to request that their interest be pooled. The primary emphasis for petition is on the working interest or mineral owner.

Fieldwide Unitization

Fieldwide unitization offers a technological advantage for the reservoir that begins to show declining pressures and production. By "secondary recovery" methods, a field operated as one entity may use gas or water injection wells to maintain pressures while additional oil is produced that might otherwise be lost to migration. The fieldwide unitization process begins with engineering and geological studies for feasibility. Unless the plan is advantageous to both the royalty owners and the working interest owners, there is no need to commence the extensive work necessary to accomplish a fieldwide unit. Two or three years of detailed, time-consuming work may be necessary.

Many states have compulsory unitization statutes that override minority nonconsenting interests for the benefit of the majority and the maximum recovery of hydrocarbons offered. The first law for purposes of gas recycling was passed in Louisiana in 1940. A more comprehensive statute was enacted in Oklahoma in 1945 and replaced in 1951 by a new law more advantageous to the royalty owner. The Oklahoma statute and an Arkansas statute (1951) have served as models for other states in allowing the respective conservation commissions to require compulsory unitization under special circumstances.[20]

A petition by the working interest owners to the regulatory commission for the unitization project must show that the project would prevent waste, avoid the drilling of unnecessary wells, protect the correlative rights of both royalty and working interest owners, and result in substantially increased ultimate recovery. The petition for the order begins with the presentation of a map and description of the area to be unitized, a list of persons owning interests in the proposed unit, a description of operations planned, and pertinent facts relating to the geological and engineering studies. The proposed unitization plan may include the preparation of either one or two unit agreements to be presented to owners for adoption. One agreement involves the working interest owners; the other is a contract that alters the terms of the oil and gas leases to allow for the fieldwide unit. Usually it is simpler to prepare the agreements into one for the operators to sign and one for the royalty owners to sign.

The statutes generally provide for the following findings:

1. Unit operation is necessary to accomplish the purposes to be ordered.
2. Proposed methods are feasible.
3. Additional recovery will exceed the costs.
4. The plan is for the common good.
5. A specified majority percentage has approved the agreements.[21]

The unitization must be substantially agreed upon by the working interest owners and operators before it can become a viable plan. These owners, together with their attorneys, engineers, and geologists, work many months to agree on specific language and operating procedures. The success of the plan also depends heavily upon the landmen charged with presenting the unit agreement to the multitude of royalty owners who must sign.

Each producing tract in the field will have a current list of persons being paid as royalty owners. The total number to be personally contacted will surely be several hundred, perhaps more than a thousand. It is essential that the landmen taking the agreement to the individuals know how to explain why it is to their advantage to participate by signing the agreement. Royalty owners' committees may be formed with the assistance of local community leaders and bankers. A common discussion for the reasons behind the secondary recovery process leads to better understanding and trust. The royalty owners may wish to hire a technical advisor or attorney to represent their views or to confirm the value of the proposed unitization. These representatives may confer many times with the technical operating committee established by the operators.

Selective production throughout the field requires that payment of royalty be based and calculated on "acre feet" of producing sand attributable to each tract, and in lieu of surface acre participation. One acre foot is deemed to be one acre in scope times one foot high. The amount of unit production to be accorded each tract is based on engineering and geological data that are appropriately weighted with regard to initial reserves, present reserves, production tests, present and projected income, porosity and permeability of the formations, crude quality, water encroachment, condensate, gas/oil ratio, and information garnered from logs, well bores, samples, and available seismic. All owners of interest are invited to ratify the unit agreement. Only working interest owners are asked to adopt the unit operating agreement, an instrument that sets out the rights and powers of the unit operator and that provides for the responsibilities and sharing of costs among the working interest parties.

A certain percentage (in accordance with state regulations) of the ownership of each tract (working interests and royalty interests) must ratify the unit agreement for that tract to qualify as participating. Those tracts that do not qualify for lack of a majority percentage in agreement may continue to be operated in accordance with their status prior to formation of the fieldwide unit. Royalty and overriding royalty owners who decline to ratify the unit are paid on actual production from their individual tracts.

State statutes vary in requirements for approving and issuing an order for the compulsory fieldwide unit. A general checklist of elements embodied in the consideration includes:

1. Description of area and detail of proposed operations
2. Allocation plan for production of oil and gas from the individual tracts
3. Financing provisions and methods of allocating costs to owners of separately owned tracts

4. Procedures and basis on which equipment and wells are to be incorporated into the unit ownership; negation of any cross-transfer of title to lands included in the unit
5. Time when unit shall become effective, and time and plan by which it may be terminated
6. Manner by which unit provisions may be modified
7. How production shall be delivered "in kind" to certain owners
8. Exemption from antitrust laws; status of the unit for tax purposes
9. Capacity of parties to sue and be sued; liabilities and nonliabilities[22]

Appropriately ratified leases within the unit are extended for the life of the entire unit, less and except "outside acreage" that might terminate by specific provision within a lease as to its nonunitized lands.

Units that embrace federal lands require the approval of the secretary through the BLM office. The federal unit agreement is complex. No specific percentage of owners is required to ratify in order for a unit to become effective, but the unit may not be approved unless the unit operator is assured control.[23]

References

[1]Howard R. Williams and Charles J. Meyers, *Oil and Gas Law* (New York: Matthew Bender, 1975), par. 901, p. 593.

[2]John S. Lowe, *Oil & Gas Law for Attorneys* (Oklahoma City: The Institute for Energy Development, 1982), p. 678.

[3]George W. Hardy, III, *Basic Louisiana Mineral Law* (Oklahoma City: The Institute for Energy Development, 1982), p. 226.

[4]Earl A. Brown, *The Law of Oil and Gas Leases* (New York: Matthew Bender, 1958), par. 15.09, p. 263.

[5]Robert E. Sullivan, *Handbook of Oil and Gas Law* (New York: Prentice-Hall, 1955), p. 153.

[6]Lowe, p. 470.

[7]Ibid., p. 492.

[8]Lewis G. Mosburg, Jr., *Problems Arising Under the Oil and Gas Lease* (Fort Worth: Institutes for Energy Development, 1974), pp. 264–265.

[9]Williams and Meyers, par. 670, pp. 394–401.

[10]Ibid., par. 905, pp. 594–595.

[11]Ibid., p. 598.

[12]Daniel J. Gamino, "Spacing, Pooling and Unitization in Oklahoma," *Landman* (January 1983), p. 35.

[13]Ibid.

[14]Eugene Kuntz, *Texas Law of Oil and Gas,* "The Texas Mineral Interest Pooling Act," by James H. Keahey (Oklahoma City: The Institute for Energy Development, 1982), p. 467.

[15]Ibid., p. 468.

[16]Ibid., p. 469.

[17]Ibid., p. 471.

[18]Ibid., p. 472.

[19]Lowe, pp. 538–539.

[20]Williams and Meyers, par. 912.1, pp. 601–602.

[21]Ibid., par. 913.4, pp. 603–604.

[22]Ibid, par. 913.6, pp. 604–605.

[23]Hugh V. Schaefer, "Unitization and Communitization," *Landman* (March 1982), pp. 10–13.

CHAPTER EIGHT

Contracts and Agreements

Contractual agreements between parties in the oil and gas industry routinely appear as letter agreements. The essential elements of a true contract should be included within an agreement:

1. Offer
2. Acceptance
3. Consideration
4. Legality of provisions
5. Legally capable parties

Oral agreements are legally binding in certain states. An agreement in writing, however, serves to protect all parties in the event of the death of witnesses present to an oral contract or in the very likely event of faulty memory. The form of the contract is not important. A memorandum may be sufficient evidence if it is executed by the party (or his legal representative) obligated to perform. The "Parol Evidence Rule" is based on the premise that an agreement in writing is the best evidence of mutual consent. This same rule of law provides that parties are bound by the contract signed, even though it varies from the agreement negotiated.[1] The statute of frauds adopted by most states requires that contracts dealing with lands be in writing. In cases of ambiguity and differing interpretation of the contract provisions, courts have tended to rule against the party who wrote the contract, for it was the preparer who had the opportunity to select the language and the other parties had relied on it.[2]

Contribution Letters

A "contribution" letter originates with a request by the first party that the second party contribute money or leasehold acreage upon the condition that the first party perform an act beneficial to the second party. The proposal most frequently refers to the first party's offer to drill a test well to a particular depth or through a specified formation. The purpose of the contribution request is to defray the expense of the test well. The second party must find that the proposed well will offer data or geological information that evaluates his own holdings in the area.

A "dry hole" contribution entails the payment by the second party of a particular sum of money upon the drilling of the test if it is found to be nonproductive or noncommercial. The importance of the language that correctly sets out the intention of the parties cannot be emphasized too strongly. In this instance, the definition of "dry hole" is critical. The payment is not due if the designated depth is not reached or if the well is completed as a producer of oil or gas.

A "bottom hole" contribution stipulates that payment be made upon reaching the agreed depth, whether the well is dry or producing. The second party contributor is accorded access "to the derrick floor" (to the well as it is being drilled). He is also entitled to copies of all logs, tests, samples, cores, and other data relating to the well. He may retain the option to take over the hole if the operator elects to plug and abandon it.

"Acreage" contributions relate to the assignment of leasehold interests to the first party upon the completion of the obligations set forth in the letter agreement. Acreage contributions may take several forms. Leasehold acreage that is owned by the second party and farmed out to the first party, offsetting or near the well drilled under the agreement, may be assigned in an undivided partial interest; or it may be assigned in full, with the second party retaining an override, production payment, carried interest, or back-in conversion to a working interest. Acreage contributions may be made on a checkerboard pattern, that is, by alternate spacing of blocks of perhaps 80 acres each, with the second party retaining every other tract.

Particular topics to be addressed in the contribution letter might include:

1. Location and manner in which the well is to be drilled, "on or before" a certain date
2. Depth to be drilled and/or formation to be tested
3. Rights of nonoperator to daily reports, notices, test information, access to well

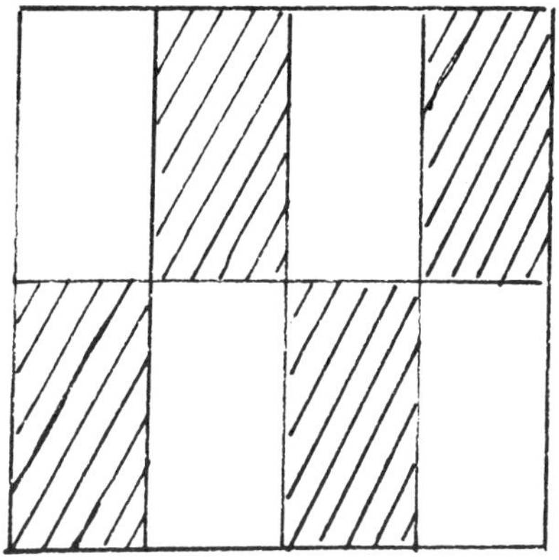

Fig. 8–1 640-acre section with an 80-acre "checkerboard"

4. Manner and time of payment to be made by second party as contributor
5. Manner and time assignment of acreage is due
6. Warranties of title (if any) on acreage to be assigned
7. Party responsible for rental payments on acreage to be contributed

Williams and Meyers outline the tax consequences to parties under contribution letter agreements:

1. Contributor
   a. Bottom hole contribution if well produces:

      Cash contribution may be capitalized as additional cost in an adjoining leasehold.

   b. Bottom hole contribution if well is dry:

      Considered expense similar to exploration costs since contributor has received geological information. Payment may be a deduction for the appropriate taxable year.

   c. Acreage contribution if well is dry:

      Deduction in taxable year, limited to cost basis in acreage.

   d. Acreage contribution if well produces:

      Cost basis of acreage capitalized as the cost of the interest received by contributor.

2. Operator
   a. Money or acreage contribution if well produces:

      If intangibles are expensed, value of acreage or cash contribution may be divided equitably between tangible and intangible costs.

      If intangibles are capitalized, cash contribution or market value of acreage would be applied to reduce capitalized depletions and depreciated costs, proportionately.

   b. If well is dry:

      If intangibles are expensed, contribution reduces ordinary loss attributable to dry hole.

      If costs are capitalized, contribution or acreage value would be applied to capitalized depletions and depreciated costs, proportionately.[3]

Farmout Agreements

A farmout agreement is a more sophisticated letter agreement that outlines the obligations of the proposed farmee (assignee) to drill one or more wells to earn acreage that will be assigned to him by farmor (assignor), upon completion of the required well(s). Leasehold owners frequently endeavor to farm out lease blocks that they do not wish to drill because of nearing expiration dates or drilling costs that the owner prefers not to assume. An operator with offsetting lease position who wishes to earn additional acreage may approach a prospective farmor with a drilling proposal. The party with the acreage to farm out stands to acquire information which may evaluate that acreage. The farmor may also save a lease from expiring by arranging to have a well drilled on the acreage or lands pooled therewith; or he may save costly rental payments with the timely drilling of a well.

The farmor generally retains either a checkerboard interest in the acreage or an undivided override, working or carried interest, or possibly an oil payment. Any of the undivided interests may be subject to a "back-in" or conversion to a full working interest status upon payout of the well or other point in time agreed upon by the parties.

The essential provisions of a farmout agreement are:

1. A complete description of all leases included under the farmout and an outline of an agreed "area of mutual interest"

(AMI) are necessary in the event that one party acquires additional leases in the vicinity. Any AMI acreage acquired after the date of the agreement might be paid for and owned proportionately by the parties to the farmout, or the lessee may be required to offer partners a proportionate interest in newly acquired acreage.

2. Copies of all leases and title information from the farmor's files are furnished to the farmee. Agreement is established as to which party is responsible for maintaining the rental payments on leases farmed out. Generally, the farmor pays rentals and bills the farmee for the entire costs.
3. A date on or before which the test well shall be commenced is established. Provisions may state that the well shall be drilled "in a good and workmanlike manner and with due diligence" to a specific depth, or to test a particular formation (called the "objective depth"). Frequently, a certain number of days is set within which the well must be completed.
4. The location of the well is agreed upon by the parties, or the location may be generally established in a quarter of a particular section, with exact location to be determined by the operator.
5. Provisions are included in the event that impenetrable formations are reached and for situations in which the prudent operator would abandon the test. This clause allows a substitute test well to satisfy the earning provisions.
6. In the event that the initial well is a dry hole, a subsequent well may be allowed at the discretion of the operator, within a specified length of time after the first well is completed as a dry hole.
7. The operator is usually charged with the entire cost and expense of drilling, completing, equipping, and/or plugging and abandoning each well. The operator is responsible for securing required permits and licenses and for complying with all laws, ordinances, and regulations governing the well operations.
8. Stipulations are set out regarding the assignment of leasehold acreage, the point at which the assignment has been earned, depths and interest to be assigned, whether renewals or extensions of the leased acreage shall be included.
9. Parties agree regarding additional wells, if any. If the farmor foresees owning a working interest in any well(s) to be drilled under the agreement, an operating agreement might be at-

tached as an exhibit, as evidence of agreement between the parties regarding future joint operations.

10. Acreage to be earned on a continuous development basis is described, and terms and manner of drilling the development wells are set out.
11. Miscellaneous provisions:
    Insurance
    Notices due
    Terms of abandonment and rights of the assignor to take over the well
    Rights of assignment and reassignment obligations
    Obligations to pay rentals and shutin royalties
    Preferential right of the assignor to purchase the product
    Provisions and penalties for forfeiture

"Time is of the essence" by the very nature of the farmout agreement, meaning timing is critical to the compliance factor. The farmee should study the proposed agreement to determine whether the obligations for earning constitute a firm covenant to comply or whether the rights to be earned are based on the condition that obligations are timely performed. The former may include specific penalties for nonperformance.

IRS Revenue Ruling 77-176 requires that the assignee must consider "excess acreage" earned under a farmout agreement as "payment for services." Value is based upon the fair market value of the excess acreage (acreage not attributable to the well drilled). The assignor must treat the same acreage as having been sold at fair market value. This "paper value" income factor may be very destructive from an economic standpoint. A common method to bypass the ruling is for the parties to the farmout to form a "tax partnership," making the terms of the tax partnership a part of the original agreement.[4] Legal counsel is advised in this matter.

Farmouts involving federal leases are subject to federal regulations.

For a particularly involved trade, stipulations of arbitration might be included in the terms.

Joint Ventures

A joint venture agreement is closely allied to the principles that govern a partnership.[5] The agreement presupposes trust among the parties to the agreement. Each has the obligation of honest and open disclosure, and one may not acquire an interest in the venture properties to the exclusion or detriment of the others.

The joint venture normally covers one undertaking or transaction. The agreement to participate is a negotiated contract constituting a fiduciary relationship. One party is generally designated the managing partner, empowered to transact the business and to sell and convey the properties within the joint venture. A joint venture may not be unilaterally abandoned without the consent of the remaining parties. The termination of a joint venture should be a conclusive act, with all assets duly assigned to the beneficial owners.

An example of a joint venture would be for acquisition of properties of any type, with each participant paying his proportionate share of costs and gaining a proportionate tax write-off or share of income and equity.

Operating Agreements

An operating agreement (or unit operating agreement) represents the terms and conditions under which the working interest owners will share costs and responsibilities for the drilling and operations of one or more wells. Each prospect, unit, or area may, and probably should, have its own operating agreement. The unleased mineral owner electing to pay his proportionate share of well costs would join in an operating agreement covering his lands.

In addition to the working interest of the lessee and/or his assigns, other working interests may be sold or "promoted" to outside investors. This is common practice in the oil industry. It allows the original investors to lessen expenses and risks, and it allows the new investor an opportunity to come into a prospect as part owner, often after monies have been expended for leases and farmouts negotiated. The promoted interest may be on any basis. "One-third for a quarter" means the investor would stand one third of the costs in return for a one-quarter net revenue interest. Promotion may be to "casing point" (the time at which election is made to set pipe and complete the well), or it may be through completion or "to the pipeline," "into the tanks," or to payout. An interest sold to a participant on a "ground floor" basis or to a participant owning an unleased mineral interest in the prospect is on an actual costs figure for the participant's proportionate share.

Prior to the commencement of a well, each participant is furnished an authority for expenditure (AFE) for approval. The AFE reflects the estimated costs that will be incurred for the particular well or workover.

For many years the AAPL Form 610 operating agreement has been used throughout the industry. Each use has seen numerous deletions and changes of its provisions to suit individual circumstances. Occasionally, a form will be prepared from scratch. The first AAPL 610 form was dated 1956. It is still in use. The 1977 revision was popular and continues to be adapted as needed. The newest revision of the model form operating agreement is the AAPL Form 610-1982 Revised (copyrighted by the American Association of Petroleum Landmen and printed by Kraftbilt Company, Tulsa, Oklahoma). This form is applicable only to nonfederal lands.

"Divided" ownerships sometimes occur within an operating area, such as within a fieldwide unit when working interests vary from tract to tract. Provisions in the unit operating agreement establish procedures whereby unit costs and participations in unit production are equitable. Most frequently, the working interests within an operating area are "undivided," with each working interest owner bearing his percentage of the whole. This factor is referred to as "participation."

All portions of the operating agreement are subject to alteration through negotiation. A general outline of the standard provisions of AAPL Form 610-1982 Revised follows:

Art. I. Definition of terminology.

Art. II. Exhibits to be attached for land descriptions; form of lease to be used in event of additional lease acquisitions; insurance coverage; accounting procedure; gas balancing agreement; tax partnership; nondiscrimination clauses; and other miscellaneous topics otherwise not included in the body of the agreement or that needed to be expanded.

Art. III. Interest of parties in the contract area, including treatment of overriding royalty or other burdens incurred or charged to one party; undisclosed or subsequently created burdens applicable to individual parties.

Art. IV. Title examination to be made on the drillsite of any proposed well, with procedure and handling of costs. Loss of title shall be borne solely by the party contributing the lease under which title has failed. Loss of a lease by failure to make a timely rental, shutin, minimum royalty, or other payment shall not create a responsibility by the negligent party for damages to other interest owners; but unlcss that lease is renewed within 90 days, contributing party is no longer credited with the interest attributed by the original contribution of that lease.

Art. V. Designation and responsibilities of the operator, with provisions for the resignation or removal of that operator. Operator may be removed if it becomes insolvent or fails or refuses to carry out its obligations. The affirmative vote of two or more nonoperators owning a majority interest is sufficient to remove one operator and select a successor. Operator shall receive bids for drilling on a competitive contract basis or use its own equipment at prevailing rates in the area. If work is performed by operator, terms and conditions shall be agreed upon in advance by participating parties.

Art. VI. Commencement date of initial test well is set, including location and depth to be drilled. Operator shall make "reasonable tests" of all formations deemed possible producing sands, unless agreement specifies only certain formations be tested. Detailed paragraphs specify procedures of operations for subsequent wells; by operations by "less than all parties"; and methods by which one party may propose a well and other parties may elect "nonconsent." One of the negotiable points of the operating agreement is the percentage of costs and expenses to be recovered by consenting parties, during which time proceeds from production attributable to nonconsenting party may be applicable to the stated percentage of recovery.

Each participant is accorded the right to take his share of production "in kind," separately disposing of the product in lieu of receiving remuneration after it is marketed. Each party has access to all information from the drilling of the well(s). Abandonment requires consent of all parties to plug. A party objecting has the right to take over the well, giving other participants the right to consent or go nonconsent.

Art. VII. Liabilities under this contract are several, not joint or collective. This means that each party is responsible only for his proportionate share of the obligations and costs. By special clause under this article, nonoperators automatically grant a lien upon extracted products and equipment in favor of operator, as security in the event of default. Operator reserves right to "call" for payments in advance to cover anticipated costs. No well shall be drilled or deepened without consent of all parties or under provisions of nonconsent clause.

Option 1: Expenditures shall be through completion and to marketing facilities.
Option 2: "Casing point" election specifies notice to partners prior to completing and equipping well to allow option to go nonconsent.

Dollar amount agreed upon up to which operator may expend money on a single project without further authorization.

Projects above the allowed expense may be solicited by operator in the form of an AFE furnished to nonoperators with a request for approval. Stipulations are agreed upon for payment of rentals, shutin well payments, minimum royalties, taxes, and other applicable assessments.

Art. VIII. Leases within contract area may not be surrendered in part or in whole without consent of all parties. If one party elects to surrender and does not have consent, he may assign that leasehold interest to the other participants, thereby being relieved of his obligations with regard to the assigned premises.

All parties are to be notified of any extensions or renewals of leases in the contract area.

Contributions of cash or acreage shall be attributable to operations under the agreement. Cash shall be remitted to operator to defray drilling costs, and acreage shall be assigned proportionately to the participants.

Insofar as laws permit, each party waives his right to partition or sever his undivided interest in the contract area. A party desiring to sell his interest in part or in whole shall notify the other participants who shall have the preferential right to purchase the interest, in the proportion each owns under the operating agreement. The preferential right to purchase does not apply to mergers, consolidations, reorganizations, or the like.

Art. IX. Internal Revenue Code election to be excluded from provisions of Subchapter "K," IRS Code of 1954, and by doing so declare no partnership association is intended by this agreement and that a partnership tax return is not necessary.

Art. X. Claims and lawsuits shall be at the joint expense.

Art. XI. Force majeure may be employed as means of extending the terms of the agreement.

Art. XII. All notices required to be in writing.

Art. XIII. Options for termination of the agreement shall be (1) for as long as any of the leases in the contract area remain in effect or (2) for as long as production or operations are in effect, or during periods of time allowed for reworking, drilling, or other operations to establish or restore production.

Art. XIV. Agreement to comply with laws and regulations.

Art. XV. Other provisions.

Under the umbrella of "other provisions" fall miscellaneous paragraphs to include fine points or provisos not included in the printed form. Conditions that may be covered are:

Consent to drilling shall not be deemed as consent to set casing. At casing point, Operator shall notify participants and each party shall notify Operator within eight (8) hours of his election to participate in completion costs ("Casing point election").

No party shall propose the drilling of more than one well at a time.

Invoices from Operator are payable within fifteen (15) days, or subject to interest at ____% per annum.

Direct charges may include legal fees up to $________.

If Operator receives 100% payment for proceeds and disburses royalties, Operator shall be entitled to charge the joint account with costs of division order title opinion and preparation of division orders.

Operator may be removed by a vote of 50% of the working interest owners.

Provisions for settling disputes: by arbitration; or by majority rule, with Operator being the deciding vote in event of a tie.

Exhibit "C" constitutes the general accounting procedure. It is a COPAS form, originally designed by the Council of Petroleum Accountants Societies of North America. The accounting procedure details which costs shall be treated as direct charges, options for overhead charges on a "fixed well" or percentage basis, and whether technical personnel shall constitute legitimate costs. The procedure may provide for amendment in the event that the rates set are found insufficient. Billing and payment methods are delineated, and rights of each party to conduct an audit are set forth.

This summary, of course, is but a thumbnail outline of a complex and detailed contract that requires considerable study and about which entire courses are taught.

Seismic Options

A seismic or "shooting" option is permission from an unleased mineral owner, generally of a large block of acreage or ranch land, to allow a company the right to conduct seismic operations upon the land. A part of the contract is the right of the optionee to select certain acreage blocks upon which to drill. Payment to the optionor of a certain number of dollars per acre for the right to shoot may be followed by payment of additional dollars per acre for "bonus" consideration (or so much cash as a penalty for nonselection) to cover a lease on selected blocks. Royalty and lease terms are negotiated in advance and are a part of the seismic option contract. Well obligations may or may not be a part of the conditions. Rights of the optionor to seismic and well information are other negotiable matters.

Multiple Use Agreements

With the advance of coal leasing across the nation, multiple use agreements between coal lessees and oil and gas lessees are increasing in number. The necessity for coexistence between the parties and for mutual responsibility and assistance parallels the rights of each party to extract or mine his product. Each lessee needs to understand the obligations incurred by the other party to develop, the timing involved, and the feasibility of interruption to the other's operations. Extensive surface areas used for coal and lignite mining may prohibit access by the oil and gas lessee to his best drillsite. Would the well have to be shut in to accommodate strip-mining procedures? Who is responsible for resetting existing pipelines? The legal conflicts require cooperation, understanding, patience, exchange of information, compromise, and regard for the rights and economic factors that motivate the opposing side.

Divison Orders

A division order is usually a brief contract for the sale of production. The division order is furnished to the royalty owner by the purchaser of the product, stating the exact royalty interest found to be attributable to the owner following title examination and division order title opinion. It describes the property and allows the royalty owner the opportunity to agree or dissent with regard to the interest set forth by his name. The division order also serves as authority for the purchase and sale of production.[6] Until such time as the title is clear or until the division order is executed by the royalty owner, the proceeds from production may be held in suspense if state law permits. A division order may contain language of ratification of a contract of sale for the product, or it may ratify a specific oil and gas lease. Such language is binding if the agreement is executed, even though the division order itself is not considered a transfer of ownership in interest.[7]

Securities Regulations

The federal government, by virtue of the Federal Securities Exchange Act of 1934 (as amended), and the various states that have adopted the Uniform Securities Act and operate under state securities commissions consider the sale of fractional interests in oil and gas leases as a sale of securities. The transfer of an entire leasehold interest would not ordinarily qualify as a security, nor would the mere execution of an oil and gas lease. However, certain contracts that "sell" or "offer to sell" undivided interests in oil and gas properties may fall under the securities or so-called "Blue Sky" laws.

The Uniform Securities Act defines a security, in part, as "any certificate of interest or participation in an oil, gas or mining title or lease, or in payments out of production under such title or lease."[8] The law requires that he who deals in the sale of securities or offers to sell securities must be a licensed dealer. The securities must be registered in advance of any offer to sell. Examples of transactions that *may* be defined as subject to the security laws are:

1. Property sales made with the implication or representation that a well will be drilled or that the land might be placed in a unit for the drilling of a well
2. Sales of undivided mineral interests
3. Assignments of small, undivided leasehold interests, with the representation that a well may be drilled in the area
4. Sales of undivided royalty interests
5. "Royalty pools" whereby an agent buys or sells various interests for a number of people holding undivided shares of the transactions

Exceptions and exemptions to the requirements to register or to be a licensed broker-dealer are allowed from state to state for limited offerings. Examples of transactions that *may* be excluded from the securities statutes are:

1. Joint venture and partnership transactions, if not offered to the public
2. "Isolated" sales
3. Personal investments
4. Limited private offerings not available to the public
5. Sales made to liquidate a debt, for the account of a mortgagee, in the course of ordinary business
6. Sales by executors, guardians, administrators, trustees, and receivers
7. Limited sales or offerings made in exchange for services or materials

Even in the absence of "willful fraud," severe criminal sanctions are imposed for violations. One engaging in transactions that may qualify as offers to sell securities is strongly advised to contact his state securities agency for full information regarding the regulations *before* any actual offering is made.

References

[1]Lewis G. Mosburg, Jr., *Contracts Used in Oil and Gas Operations* (Fort Worth: Institutes for Energy Development, 1982), p. 32.

[2]Ibid., p. 23.

[3]Howard R. Williams and Charles J. Meyers, *Oil and Gas Law,* abr. ed. (New York: Matthew Bender, 1975), par. 431.2, pp. 240–241.

[4]Mosburg, p. 45.

[5]Williams and Meyers, par. 437, p. 249.

[6]Mosburg, p. 57.

[7]Williams and Meyers, par. 707–708, pp. 452–453.

[8]Frank K. Young, "Landmen and Securities," *Landman* (December 1981), pp. 4–5.

CHAPTER NINE

Lease Maintenance

Lease maintenance is a broader responsibility than the regular timely payment of delay rentals. The lease records department of a company involves itself with proper record keeping and call-up systems to allow a period of time to review the lease block and the terms of each lease prior to rental payment.

Computer data systems for lease records depend upon accurate input. Data call-up is generally provided three to six months before the obligation date and then monthly until the date arrives. Obligations noted on the call-up system—whether on a computer or by an in-house method—include not only delay rental dates but shutin well payments, well obligation dates, reassignment obligations, expiration dates, and notices due to partners. "Setting up" a new lease requires a conscientious study and notation of all provisions.

Critical to lease maintenance is the timely compliance with all provisions in the instrument, from offset well obligations to implied covenants to develop, from the saving clause to payments of minimum royalty. This chapter covers the basic resources and practices available for maintaining a lease in full force and effect.

Timely Payments

Leases that provide for delay rental payments to defer commencement of operations allow the lessee to tender the delay rental prior to the anniversary date of the lease, each year within the primary term. The anniversary date is the effective date of the lease, not the date of its execution. The "paid up" lease has no provision for delay rentals, for it is "paid up" until the end of the primary term. Many lease forms, particularly in the Midwest, provide for semiannual or quarterly rental payments.

Two general wordings of rental clauses have been discussed in Chapter 5. Again, these clauses are "unless" (whereby *unless* operations are commenced or rentals paid, the lease automatically terminates) and "or" (operations may be commenced *or* rentals paid *or* the lease must be surrendered; lessor may recover damages unless one of the options is effected). California leases adhere almost exclusively to the "or" clause, which imposes a duty to perform upon the lessee.[1] Most other states use both types of provisions. The "unless" clause imposes no activity on the lessee if he does not wish to maintain the lease.

John S. Lowe recites four instances in which late payments might serve to preserve a lease:

Late payment is accepted by lessor
Failure attributed to third party (e.g., post office or bank)
Lessor causes payment to be late or underpaid
Lessor fails to notify lessee of an inadequate or incorrect payment[2]

Lowe states that the "principle of cooperation" under a contract does not allow the lessor the privilege of holding a tendered rental while he determines whether it might be more profitable to declare the lease void.[3] The lessee may also be excused for improper payment of rental if payment is based on an ambiguous instrument or conveyance. The classic case cited by Lowe to illlustrate his theories is Humble Oil & Refining Co. v. Harrison, 146 Tex. 216, 205 S.W.2d 355.

Shutin well payments are most frequently designated in the lease language as payments in lieu of royalty from production. Nonetheless, shutin payments are treated with kid gloves by most operator/lessees, and efforts are conscientiously made to make timely shutin payments to avoid what might be a disastrous loss of a lease capable of producing. "Production" required under the terms of the lease to extend the term is literally interpreted in the majority of states. Oklahoma has ruled that the "capability of production" is sufficient to maintain the lease, Gard v. Kaiser, 582 P.2d 1311 (Okla. S.Ct. 1980).[4] Therefore, in Oklahoma, the inadvertent failure to pay a shutin royalty does not terminate the lease.

Shutin payments apply only to a well capable of production. The well that ceases to produce and that must be worked over would place the lease within the "savings" or "cessation of production" stipulations.

Regardless of court cases granting revival of the lease terms in the event of late or insufficient payment, the general ruling is that improper payment or late payment is tantamount to no payment.

Other payment obligations under a lease may include minimum rentals that continue even after discovery of production and payment of royalties; minimum royalty payments; rentals on "outside" or nonproducing acreage; compensatory royalty in lieu of the drilling of an offset well.

What is the recourse of the lessor in the event that royalties are not paid or are held for undue periods of time? In the absence of an express provision requiring payment of royalties within a specified time or forfeiture of the lease, limited damages are available through court action. Cancellation of the lease for failure to pay royalties was an available recourse in Louisiana prior to the enactment of the Louisiana Mineral Code of 1975. Under the Mineral Code, the lessor must now show fraud or misconduct on the part of the lessee to sue for cancellation of the lease. Otherwise, he may seek to recover double the amount due as damages, plus attorney's fees.[5]

Operations or Production

The "dry hole" clause in a lease may provide for the suspension of delay rentals during the year in which the dry hole was drilled. Resumption of rentals is generally due for the year following the drilling of the dry hole and for the remaining years of the primary term, barring other operations. Dry hole provisions are notorious for their variance in wording; many questions may be left unanswered regarding the rights of the lessee. How long may rentals be suspended following the drilling of the dry hole? If the operator elects to drill another well, is the provision clear as to how long he may wait to commence additional operations? In the absence of specific time limitations, the lessee would likely be required to commence operations within a reasonable period of time, if the lease continues within the primary term. A fair consideration for both parties seems to be that operations be resumed within 90 days or that rentals be paid on the next anniversary date following 60 or 90 days after the well is completed as a dry hole.

What is the definition of a dry hole? In the absence of definition within the lease, a dry hole is generally construed to be a well that will not produce in paying quantities. How deep must the well be drilled to be considered a dry hole? The rule of thumb is that a well must be drilled to the depth at which oil or gas is usually discovered in the area.[6] If there are two or more producing horizons in the vicinity, is a well drilled only to the upper horizon considered a dry hole? What if the well is a wildcat—drilled in an untested area—and the possible producing depth is unknown? The likelihood of differing interpretations would be eliminated if the parties to the lease inserted a definition of "dry hole."

The habendum clause usually states the primary term for which the lease will be effective "and as long thereafter as production continues in paying quantities." A well drilling across the end of the primary term would not hold the lease because production had not been obtained at that point. (Again, the exception might lie in Oklahoma, where the courts have permitted the lessee to complete operations that were commenced before the end of the primary term.) A continuous operations clause within the lease grants the lessee this option:

> Notwithstanding anything in this lease to the contrary, it is expressly understood and agreed that if Lessee shall commence operations for the drilling of a well at any time while this lease remains in full force and effect, this lease shall remain in effect and its terms shall continue for as long as operations are continued; and if production is discovered, for as long as such production continues.

Present-day lease forms are likely to include a "cessation of production" or so-called "saving clause," which allows the lessee 30 to 60 or 90 days following cessation of production in which to (1) resume the payment of rentals on the next anniversary date ensuing or (2) commence reworking or drilling operations in an attempt to restore production. Cessation of production is generally construed to mean complete cessation of production, not merely cessation of production in paying quantities.[7] If the lease contains a shutin gas payment provision and the well has been a producer of gas, the lessee might fall back on the shutin payment to maintain the lease while reworking operations are planned. This course would not be available if the shutin payment could be made only in the event of no market for the gas. Kansas courts have held that if a lease contains a drilling operations provision, the lease term is extended for "as long as the lessee diligently pursues putting it into production."[8] John S. Lowe cites Sword v. Raines, 575 F.2d 810 (10th Cir. 1978) as an example of the Kansas decision, in which more than eight months expired before a pipeline connection could be made for a gas well. The ruling was in favor of the defendant lessee.

Force Majeure

The burden of proof is upon the lessee to establish that a force majeure cause interrupted his performance under the terms of a lease that contains this clause. Force majeure is defined by *Black's Law Dictionary* and by *Webster's Seventh New Collegiate Dictionary* as a "superior or irresistible force." Such conditions take into account governmental authority and regulations that prevent continuing performance; acts of God; war; weather occurrences; or other suspensions of performance or production out of the control of the lessee. Litigation affecting the title to the land might be considered force majeure.[9] Recent force majeure clauses have included the inability to obtain necessary equipment or materials for operations.

Under this umbrella of protection, the lessee may be allowed a suspension of the covenants and obligations imposed, with an extension of time allowed equal to the period of suspension. The lessee is held not liable for damages in most cases, but he may be required to continue payments of rentals or royalties during the period that force majeure is in effect.

References

[1]Eugene Kuntz, *Texas Law of Oil and Gas,* "The Drilling Delay Rental Clause," by John S. Lowe (Oklahoma City: The Institute for Energy Development, 1982), p. 243.

[2]Ibid., pp. 258–259.

[3]Ibid., pp. 259–260.

[4]John S. Lowe, *Oil & Gas Law for Attorneys* (Oklahoma City: The Institute for Energy Development, 1982), p. 332.

[5]Ibid., p. 346.

[6]Howard R. Williams and Charles J. Meyers, *Oil and Gas Law,* abr. ed. (New York: Matthew Bender, 1975), par. 614.1, p. 328.

[7]Ibid., par. 616.1, p. 332.

[8]Lowe, p. 328.

[9]Williams and Meyers, par. 683.1, p. 423.

CHAPTER TEN

Land Measures and Descriptions

"A survey does not merely identify boundaries, it creates them."[1] The word "survey" may be defined as a noun or as a verb. As a noun it means a system of measuring to determine the contents of a tract of land. As a verb it means to ascertain corners and boundaries, with the aid of distance and direction measurements.

The original 13 states, along with Kentucky, Tennessee, West Virginia, Vermont, and Maine, were mapped by an early system called the "colonial survey."[2] This method designated townships six miles square. Individuals were granted interior parcels of irregular configurations that tended to conform to the natural features of the land.

The National Land Act of 1785 established the rectangular or governmental survey system (also called congressional townships) by which all United States public lands would thereafter be surveyed prior to any sale or disposition. The rectangular survey is based on six-mile-square townships. An area six townships square is called a "check." The township is comprised of 36 sections, each one mile square.

The Land Office in Washington, D.C., located north-south lines termed "meridians." Crossing the meridians in east-west directions are "base lines." The township survey begins at a point on which the meridian and base line meet. Irregular sections on the north and west sides of townships adjust the measurements for earth curvature.

Township lines are six miles apart and run east-west. Range lines, six miles apart, extend north-south and are numbered from the meridian line, beginning with the number one (1). The township

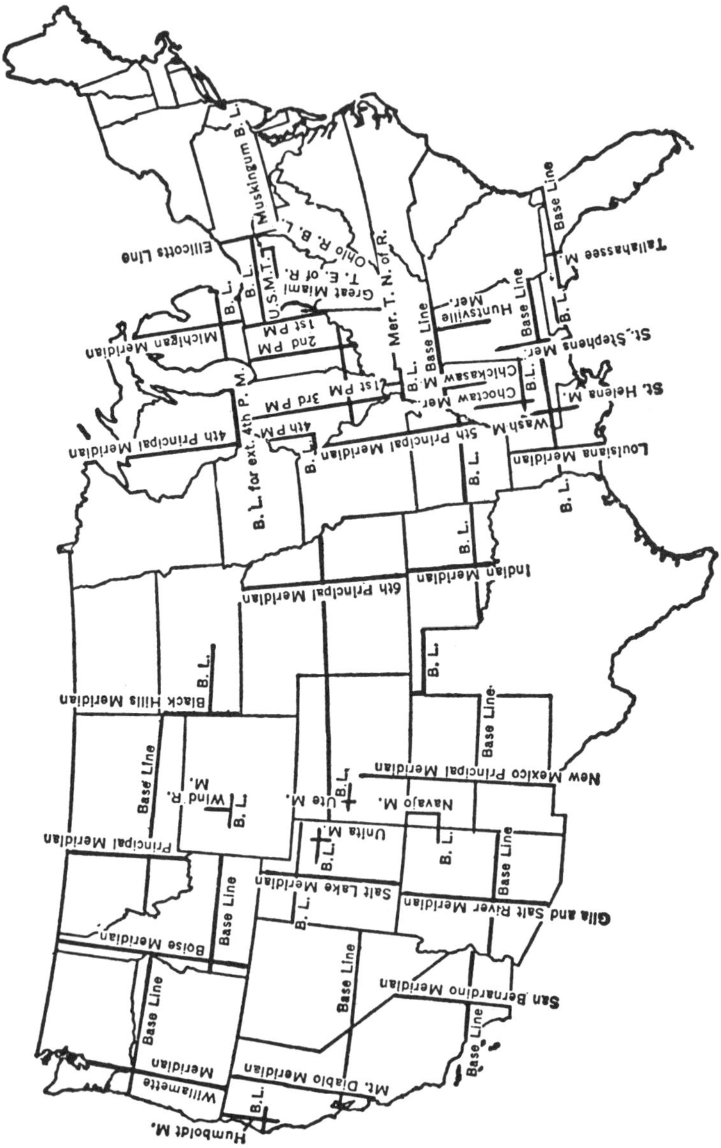

Fig. 10–1 Map of prime meridians and their base lines within the United States (Alfred A. Ring, Nelson L. North, *Real Estate Principles and Practices*, 6th ed. p. 93. Reprinted by permission of Prentice-Hall, Inc., Englewood Cliffs, N.J.)

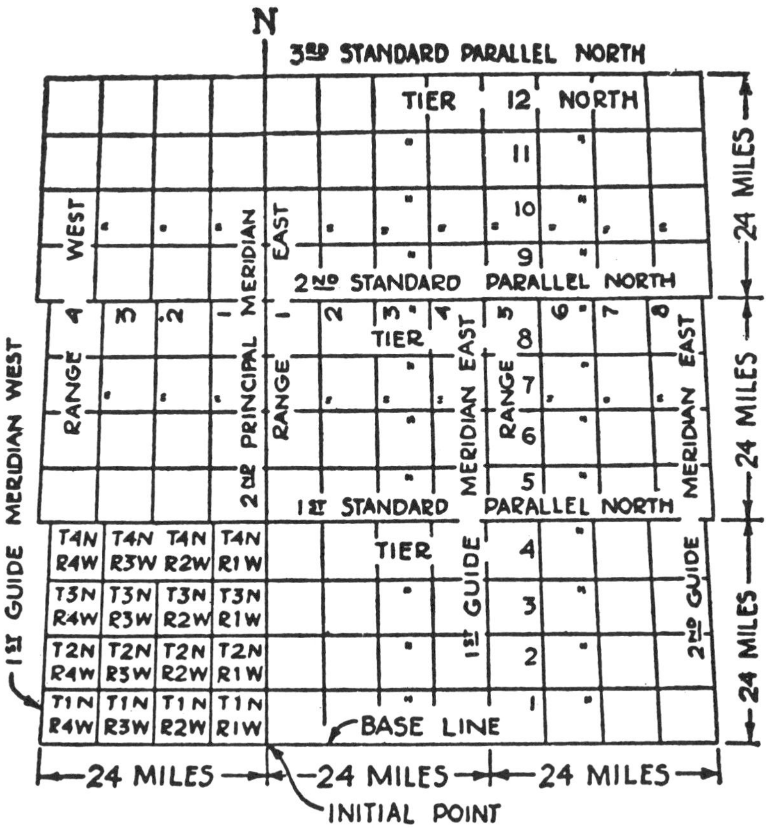

Fig. 10–2 Township and range numbering system (Alfred A. Ring, Nelson L. North, *Real Estate Principles and Practices,* 6th ed. p. 94. Reprinted by permission of Prentice-Hall, Inc., Englewood Cliffs, N.J.)

lines that cut the range lines into squares are numbered from the base line, beginning with number one (1).

Example: A township situated to the immediate north of the base line and two townships west of the meridian would be designated Township 1 North, Range 2 West (or T-1-N, R-2-W or T1N-R2W).

Each township is measured into 36 sections, one square mile each (there being 640 acres per section). Sections in a township are numbered 1 through 36, beginning with the northeast corner, westerly through section 6. Then they drop south one section,

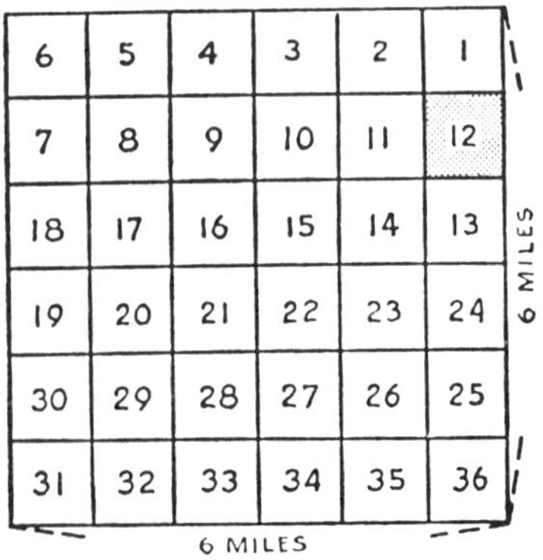

Fig. 10–3 Section number system (Alfred A. Ring, Nelson L. North, *Real Estate Principles and Practices,* 6th ed. p. 94. Reprinted by permission of Prentice-Hall, Inc., Englewood Cliffs, N.J.)

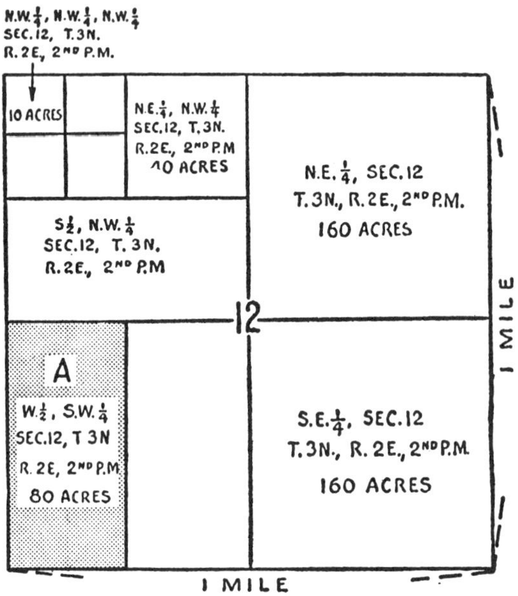

Fig. 10–4 Rectangular survey section partition (Alfred A. Ring, Nelson L. North, *Real Estate Principles and Practices,* 6th. ed. p. 95. Reprinted by permission of Prentice-Hall, Inc., Englewood Cliffs, N.J.)

beginning with section 7, and proceed in an easterly direction through section 12 (which lies immediately south of section 1). The numbers continue easterly, then westerly, ending with section 36 in the southeast corner. Each section is divided into quarters:

NE¼ NW¼ SW¼ SE¼ (160 acres each)

These quarters are in turn quartered:

NE¼ of the NW¼ (40 acres)

And those, too, are quartered:

NE¼ NW¼ NW¼ (10 acres)

To locate a tract designated as the SW¼ of the NE¼ of the NE¼, it is easier to read the description backward: the parcel is in the NE¼ of the section, thence in the NE¼ thereof, and thence in the SW¼ of the NE¼ of the NE¼. This tract would be 10 acres.

A description that included the SW¼, S½NW¼ would be:

SW¼	160 acres
S½NW¼	80 acres
	240 acres

The location of the comma in the description sets off one tract from the other. Tracts are described by section, township, range, county, and state.

The outer continental shelf is surveyed into areas, blocks, and tracts, with measurements based on Lambert plane coordinates.

The old "Virginia system" is widely used in the South, in some Atlantic Coast states (e.g., Virginia, the Carolinas, and Georgia), and in Texas. It is a method of description by metes (measurements) and bounds (boundaries), using natural landmarks and monuments. A fixed starting point for the survey might be an oak tree marked with an "X," with calls of so many feet or other length of measurement and degrees of angle to the next marker and continuing until the description "closes." It is easy to understand why so many boundary disputes have arisen from metes and bounds descriptions. Many created vacant spaces (vacancies) between tracts or overlap. Nevertheless, the landman or landowner can locate his tract on a properly scaled map with an engineering (Allen) scale, by following the metes and bounds surveyed. Brown relates a metes and bounds description in a New Mexico lease calling for a certain distance "down the canyon to the place where the men were killed."[3] The location related to a well-known landmark in the area and was deemed sufficient to serve as an accurate legal description.

Streams and bodies of water serve as legal boundaries. The meandering of a river or stream may alter the riparian configuration by accretion, reliction, or erosion of the soil affecting the land. Avulsion, a sudden and noticeable alteration in the course of the water, does not change the boundary, which would continue to the midpoint or center of the old riverbed.

Riverbeds of navigable waters generally belong to the states. Private ownership of adjoining lands terminate at the high-water mark.

The Spanish heritage of California and the southwestern United States is depicted in the continuing use of the Spanish *vara,* a term meaning "yard." The exact length of the vara differs in the several states from 32.8748 inches to 35 inches. It is generally established to be 33.33 inches. Similarly, in the southern portion of Louisiana, the French *arpent* remains in use. Its length varies from parish to parish but in general is counted to be 191.833 feet.

Subdivision plats, lots, and block references constitute legal descriptions of city parcels. Each includes the name of the city or town, county or parish, and state.

Common units of measure are referenced in the tables that follow. The reader is referred to the *AAPL Guide for Landmen,* rev. ed. 1980, by Leslie Moses, for detailed tables of measurement and conversion.

References

[1]*63 American Jurisprudence,* 2d, ser. (San Francisco: The Lawyers Co-Operative Publishing Co., 1980), par. 39; State v. Allen, 274 Ala. 600.

[2]Lewis G. Mosburg, Jr., *Landman's Handbook on Petroleum Land Titles* (Fort Worth: Institutes for Energy Development, 1976), p. 29.

[3]Earl A. Brown, *Law of Oil and Gas Leases* (New York: Matthew Bender, 1958), par. 4.01, p. 42.

Acreage

1 acre	=	43,560 square feet 4,840 square yards 160 square rods 10 square chains 208.71 feet square
1 township	=	36 square miles 6 miles square 23,040 acres
1 section	=	1 square mile 640 acres
½ section	=	320 acres
¼ section	=	160 acres
⅛ section	=	80 acres
1/16 section	=	40 acres
1/64 section	=	10 acres

Linear measures

1 mile	=	1,760 yards 5,280 feet 320 rods 80 chains
1 chain	=	66 feet 100 links or 4 rods
1 rod (pole)	=	5.5 yards 16.5 feet 25.0 links
1 yard	=	3 feet 36 inches
1 link	=	7.92 inches
1 inch	=	0.0833 feet

Spanish land measures

1 labor	=	177.136 acres	=	1,000,000 square varas
1 league	=	4,428.4 acres	=	25,000,000 square varas 25 labors
1 league and 1 labor	=	4,605.54 acres	=	26,000,000 square varas
1 sitio (sitio de ganado mayor)	=	1 league 4,428.4 acres		
1 sitio de ganado menor	=	1,968.18 acres		
1 caballeria	=	108 acres		
1 vara	=	$33^{1}/_{3}$ inches or 2.778 feet		
36 varas	=	100 feet		
108 varas	=	100 yards		
5,645.376 square varas	=	1 acre		

Conversions

Varas to feet		Divide by 0.36
Varas to yards		Divide by 1.08
Varas to rods		Divide by 5.94
Varas to chains		Divide by 23.76
Square feet to square varas		Divide by 7.716
Square varas to acres		Divide by 5,645.37
Square feet to acres		Divide by 43,560
1 foot	=	0.35 vara
5 feet	=	1.80 varas
10 feet	=	3.60 varas
25 feet	=	9.00 varas

French land measures

1	arpent	=	191.833 feet
1	square arpent	=	0.84628 feet
1.1834	square arpents	=	1 acre

Code of Ethics*

Bylaws
American Association of Petroleum Landmen

Article IV

The Code of Ethics shall be the basis of conduct, business principles and ideals for the members of the American Association of Petroleum Landmen; and it shall be understood that conduct of any member of the Association inconsistent with the provisions set forth in this Article shall be considered unethical and said individual's membership status shall be subject to review for possible disciplinary action as prescribed in Article XVI of these Bylaws.

In the area of human endeavor involving trading under competitive conditions, ethical standards for fair and honest dealing can be made increasingly meaningful by an association organized and dedicated not only to the definition, maintenance and enforcement of such standards, but to the improvement and education of its members. Such is the objective of the American Association of Petroleum Landmen and such is its public trust.

Section 1. It shall be the duty of the landman at all times to promote and, in a fair and honest manner, represent the industry to the public at large with the view of establishing and maintaining good will between the industry and the public.

The landman, in his dealings with landowners and others outside the industry, shall conduct himself in a manner consistent with fairness and honesty, such as to maintain the respect of the public.

---

*Reprinted by permission from the American Association of Petroleum Landmen, Fort Worth, Texas, Membership Directory 1982–83, pp. 401–402.

Section 2. Competition among those engaged in the mineral and energy industries shall be kept at a high level with careful adherence to established rules of honest and courtesy.

A landman shall not betray his employer's (or client's) trust by directly turning confidential information to personal gain.

The landman shall exercise the utmost good faith and loyalty to his employer (or client) and shall not act adversely or engage in any enterprise in conflict with the interest of his employer (or client).

The landman shall represent to others his area of expertise and shall not represent himself to be skilled in professional areas in which he is not professionally qualified.

Glossary

A

Abstract Bound compilation of all instruments of record affecting the title to lands specifically described.

Adverse possession Ownership claimed against the holder of record title by hostile, notorious, and open possession.

Affiant One who makes or executes an affidavit.

Affidavit Sworn statement of facts.

After-acquired title Interest acquired by grantor/lessor subsequent to execution of a conveyance or lease.

Alluvial Buildup of soil by deposit from moving waters.

Arpent French land measure, 191.8 feet, more or less.

Assignee Party receiving an assignment of interest.

Assignment Conveyance to transfer interest.

Assignor Party making or granting an assignment.

B

Back-in Reversionary interest, generally to convert an override or carried interest to a full working interest at a designated point in operations or at payout.

Beneficial interest Ownership by virtue of an agreement, contract, or trust as opposed to record or legal outright ownership and control.

Bonus Cash consideration for initial purchase of an oil and gas lease.

Bottom-hole contribution Cash payment in consideration of agreement and performance in the drilling of a well to a specified depth.

Burdens Lessor royalties, overriding royalties, production payments, carried interests, and back-ins to which a leasehold working interest is subject.

C

Carried interest Working interest on which costs are paid out of revenues attributable to the carried interest, with no personal obligation for payment of costs by the owner of the interest. Generally converts to full working interest at casing point or at payout.

Communitize Pool into one entity, as small tracts into one unit.

Community property Equal ownership between husband and wife of properties acquired during marriage; excludes "separate" properties owned prior to marriage and properties inherited.

Compensatory royalty Payment in lieu of production royalty.

Completed/completion Point at which a well is fully drilled; may refer to dry hole or to a producing well that is "completed" by installation and setting of pipe and equipment to ready the well for production.

Compulsory pooling Involuntary, or forced by legal statute, to consider all interests as proportionate parts of one entity for exploration and production purposes.

Correlative rights Mutual or reciprocal existence and obligation.

Covenant Firm promise to perform.

Curative Title requirements to remove objections or clouds on record title.

Curtesy Life estate created for a widower if issue who will inherit survive; not effective in community property states.

D

Dedication Formal conveyance of private land for public use.

Deed Written conveyance instrument for purpose of transferring title.

Delay rental Annual or more frequent payment due under terms of oil and gas lease for privilege of delaying the commencement of drilling operations.

Division order Contract between purchaser and royalty owner affirming percentage of proceeds due to royalty owner and allowing sales of products attributable to royalty owner.

Dower Life estate created for a widow if issue who will inherit survive; not effective in community property states.

E

Easement Written permission to use the surface of described lands for designated purpose. "Implied" easement constitutes unwritten right of ingress and egress.

Eminent domain Taking of private lands for public use by a governmental authority with just compensation to the private owner.

Entirety One entity, as opposed to a part or moiety, when the whole is owned by two or more parties.

Equitable servitude Deed or zoning restrictions imposed on surface use.

Executive right Legal right to sign an oil and gas lease.

F

Farmee Party receiving obligations and benefits of a farmout agreement.

Farmor Party granting a farmout contract.

Farmout Contract allowing farmee to earn interest in leasehold acreage of farmor upon performance of obligations to drill.

Fee Record ownership of surface or minerals.

Fee simple Absolute ownership of interest in land, of surface and all minerals.

Finders fee Cash consideration, royalty, overriding royalty, working interest, or other interest due and payable to third party upon the acceptance and consummation of a farmout contract, drilling deal, or other type of investment program between first and second parties, brought about by third party.

Force majeure Acts of God and war and weather occurrences beyond the control of the parties to a contract.

Freestone rider Special provision to oil and gas lease, provides for termination of acreage not included in a unit at end of primary term of lease.

G

Grantee Recipient of a conveyance.

Grantor Party making a conveyance in favor of another.

H

Habendum Clause in deed or lease that defines extent of ownership conveyed to grantee.

Headright Undivided ownership under the whole; a prorata beneficial interest (Indian Allotment Act of 1906, 34 Stat. 539).

Headright certificate In Republic of Texas between October 1, 1837, and January 1, 1840, a grant of 640 acres to head of family, subject to residency requirements.

J

Joint tenancy (JTROS) Concurrent ownership with right of survivorship.

Jurat Certificate of sworn statement taken before a notary public or other official.

L

Labor Spanish land measure, 177 acres, more or less.

League Spanish land measure, 4,428 acres, more or less.

Lessee Grantee under terms of oil and gas lease or other lease.

Lesser estates/Lesser interest clause Proportionate reduction of payments under oil and gas lease in ratio of fractional interest owned by lessor to the whole.

Lessor Grantor of oil and gas or other lease.

Life estate/Life tenant Beneficial ownership for lifetime; interest reverts to remainderman upon death of life tenant.

Limitation title Indefeasible ownership acquired by continuous use and possession over a period of time exceeding the statute of limitations.

M

Minerals Inert substances in soil or rocks, deposited through natural causes; fossil or inorganic.

Muniment of title Instrument filed to serve as a link in the chain of title.

N

Net profits Proceeds from production attributable to lessee after all costs of drilling and development are satisfied.

Net revenue interest (NRI) Percentage of production due to working interest owners after deductions of lessor royalty, overriding royalty, and other leasehold burdens from 100%.

Nonparticipating royalty (NPRI) Nonexecutory interest sharing in proceeds of production out of lessor royalty; does not share in bonus or rental payments but may be entitled to share of shutin royalty.

O

Offset well Implied covenant under oil and gas lease that requires lessee to drill to prevent drainage from well on adjoining lands, provided offset well would likely show a profit. May be required by express covenant.

Operating agreement Contract between working interest owners designating terms and accounting procedures under which an area of interest will be developed.

Operator Entity or individual responsible for the drilling operations of a well and maintenance of the lease property.

"Other minerals" In oil and gas lease, generally all minerals other than those of hydrocarbon nature.

Overriding royalty Royalty over and above the lessor royalty, free and clear of costs other than attributable taxes.

P

Parol Evidenced by instrument in writing.

Participation Divided or undivided fractional working interest under an operating agreement.

Partition Vertical severance to segregate undivided interests into separate surface tracts; horizontal severance as to specified formations.

Pool Unitize into one tract for drilling and production purposes.

Power of attorney Instrument granting signatory powers to an agent and attorney-in-fact. May be general or specific.

Preferential right to purchase Right of first refusal; first "call" on gas or oil produced. Right to purchase product at prevailing market rate; or right to purchase interest of another party desiring to sell, or to waive such right before offer is made to another.

Primary term Basic time period under which an oil and gas lease is effective.

Production payment Contract lump-sum payment to be made out of proceeds from production over a period of time.

Pugh clause Special provision under oil and gas lease to pay delay rentals on acreage not included in a unit for designated number of years following end of primary term.

Q

Quitclaim deed Disclaimer of interest.

R

Ratification Agreement and adoption of terms of another instrument specifically described as if ratifying party had executed the original instrument.

Release Instrument of termination.

Remainderman Owner of corpus after death of life tenant.

Reservation Right retained by grantor in a conveyance; or lands set aside by governmental authority for use and occupation of Indian tribe.

Resolution Nonrecorded instrument prepared by board of directors of a corporation empowering an officer to act on its behalf.

Reversionary Future estate or interest restored after a present possessor dies; future interest created by conversion of a present interest.

Riparian Lands fronting the waters of a stream or river.

Royalty Interest created by existence of oil and gas lease; proportionate share of production from sale of minerals, free of costs other than applicable taxes.

Royalty acre Full leasehold royalty on one acre of land times one-eighth (12.5%) lessor royalty.

Rule against perpetuities Interest in property must vest not later than 21 years after a life that is in existence at time interest is created.

Rule of capture Mineral owner is allowed to produce all he can "capture" by drilling from his own lands, in recognition of migratory nature of oil and gas.

S

Secondary recovery Water or gas injection or other methods to stimulate depleting reservoirs.

Seismic Electric or shock impulses that translate subsurface contours to line images.

Separate property Property acquired or owned prior to marriage or by inheritance by one spouse exclusive of the other.

Servitude A restricted right, easement, or burden on property. In Louisiana, a severed mineral interest.

Sheriff's deed Conveyance under terms of bankruptcy proceedings.

Software Computer printout of leasehold or other data.

Sovereignty Title in original governmental authority.

Special warranty Terms of limitation regarding warranty of title.

Subordination Placement in descending or inferior legal status.

T

Take "in kind" To take one's proportionate share of production and separately sell or dispose of it rather than to accept payment in funds.

Tenancy by the entirety Concurrent ownership betwen husband and wife with right of survivorship.

Tenants in common Concurrent ownership without right of survivorship.

Tight hole A drilling well on which operator is maintaining all data in confidentiality.

Title opinion Attorney's written report of examination of record title.

Top lease Lease taken to be effective on expiration of present lease.

Torrens system Means of title registration used by only a few states, e.g., Illinois. Court directs issuance of title certificate upon application by landowner.

Trespass to try title (TTT) Legal actions to settle a controversy between rival claims or to quiet title.

Trustee Party acting in a fiduciary capacity for another.

U

Undivided interest Fractional interest under the whole.

Unit The pooling of two or more tracts for production.

Usufruct In Louisiana, a life tenancy.

V

Vara Spanish measure, $33\frac{1}{3}$ inches, more or less.

W

Warranty Guarantee.

Wildcat Well drilled in an untested area.

Working interest Lessee interest under oil and gas lease, obligated to pay costs of drilling and development.

Bibliography

Primary Sources

A.A.P.L. Educational Programs, Inc. "Advanced Land Practices." Fort Worth: American Association of Petroleum Landmen, 1982.

American Jurisprudence. 2d ser. San Francisco: The Lawyers Co-Operative Publishing Co., 1980.

"Annual Survey of Texas Law, Mineral Resources." *Southwestern Law Journal,* vol. 36, no. 1. 1982.

Black, Henry Campbell. *Black's Law Dictionary.* 4th ed. rev. St. Paul, Minn: West Publishing Co., 1968.

Brown, Earl A. *The Law of Oil and Gas Leases.* New York: Matthew Bender, 1958.

Caskey, C. J. "610 Model Form Revisions." *Landman,* June 1982, p. 6.

Corpus Juris Secundum. Brooklyn, N.Y.: American Law Book Company, 1944.

Dobkins, Betty Eakle. *The Spanish Element in Texas Water Law.* Austin: University of Texas Press, 1959.

Ernest, Max H., III. "Top Leasing—Legality v. Morality." Paper presented at the 26th annual Rocky Mountain Mineral Institute, Sun Valley, Idaho, July 1980.

Field, Truman. "Units and the Division Order." *Landman,* January 1981, pp. 52–55.

Gamino, Daniel J. "Spacing, Pooling and Unitization in Oklahoma." *Landman,* January 1983, pp. 35–40.

Gibson, Charles. *Spain in America.* New York: Harper & Row, 1966.

Hardy, George W., III. *Basic Louisiana Mineral Law.* Oklahoma City: The Institute for Energy Development, 1982.

Hemingway Richard W. *The Law of Oil and Gas.* St. Paul, Minn.: West Publishing Co., 1971.

Israel, Daniel H. "Indian Lands, Current Legal Issues Affecting Oil & Gas Development on Indian Reservations." *Landman,* November 1982, pp. 50–60.

Kuntz, Eugene. *Texas Law of Oil and Gas.* Oklahoma City: The Institute for Energy Development, 1982.

Lowe, John S. *Oil & Gas Law for Attorneys.* Oklahoma City: The Institute for Energy Development, 1982.

Mosburg, Lewis G., Jr. *Contracts Used in Oil and Gas Operations.* Oklahoma City: The Institute for Energy Development 1982.

———. *Landman's Handbook on Petroleum Land Titles.* Fort Worth: Institutes for Energy Development, 1976.

———. *Problems Arising Under the Oil and Gas Lease.* Fort Worth: Institutes for Energy Development, 1974.

Moses, Leslie. *The AAPL Guide for Landmen.* Rev. ed. by Milton L. Brand. Tulsa: Kraftbilt, 1980.

Ring, Alfred A., and Nelson L. North. *Real Estate Principles and Practices.* 6th ed. Englewood Cliffs, N.J.: Prentice-Hall, 1967.

Schaefer, Hugh V. "Unitization and Communitization." *Landman,* March 1982, pp. 10–21.

Smith, Chester. *Smith's Review, Legal Gem Series, Real and Personal Property, Conveyancing and Future Interests.* Rev. ed. by Howard M. Rossen and Wilton S. Sogg. St. Paul, Minn.: West Publishing Co., 1972.

Stitt, Edwin. "Eastern Overthrust Belt." *Landman,* June 1982, pp. 29–30.

Sullivan, Robert E. *Handbook of Oil and Gas Law.* Englewood Cliffs, N.J.: Prentice–Hall, 1955.

Tait, Samuel W., Jr. *The Wildcatters.* Princeton, N.J.: Princeton University Press, 1946.

Williams, Howard R., and Charles J. Meyers. *Oil and Gas Law.* Abr. ed. New York: Matthew-Bender, 1975.

Williamson, Harold F., and Arnold R. Daum. *The American Petroleum Industry.* Evanston, Ill.: Northwestern University Press, 1959.

Young, Frank K. "Landmen and Securities." *Landman,* December 1981, pp. 4–5.

Secondary Sources

Avina, Rose Hollenbaugh. "Spanish and Mexican Land Grants in California." Thesis, University of California, 1932.

Baade, Hans W. "The Formalities of Private Real Estate Transactions in Spanish North America." *Louisiana Law Review* vol. 38, no. 3, (1978).

Bourne, Edward Gaylord. *Spain in America.* New York: Harper & Row, 1904.

Boyle, Gary. "Indian Lands," *Landman,* October 1981.

Burns, James MacGregor, and Jack Walter Peltason. *Government by the People.* Englewood Cliffs, N.J.: Prentice-Hall, 1972.

Cohen, Morris L. *Legal Research.* Nutshell Series. St. Paul, Minn.: West Publishing Co., 1981.

Giddens, Paul H. *Early Days in Oil.* Princeton, N.J.: Princeton University Press, 1948.

Morrow, William H. *Spanish and Mexican Private Land Grants.* San Francisco: Bancroft-Whitney Co., 1923.

Taylor, Virginia H. *Index to Spanish and Mexican Land Grants.* Austin: General Land Office of Texas, 1976.

Warner, C. A. *Texas Oil and Gas Since 1543.* Houston: Gulf Publishing Co., 1939.

Index

P

Q

R

S